EXPERT ANDROID® STUDIO

W9-CIC-334

EXPERT

Android® Studio

EXPERT

Android® Studio

Murat Yener

Onur Dundar

Expert Android® Studio

Published by
John Wiley & Sons, Inc.
10475 Crosspoint Boulevard
Indianapolis, IN 46256
www.wiley.com

Copyright © 2016 by John Wiley & Sons, Inc., Indianapolis, Indiana

Published by John Wiley & Sons, Inc., Indianapolis, Indiana

Published simultaneously in Canada

ISBN: 978-1-119-08925-4
ISBN: 978-1-119-11071-2 (ebk)
ISBN: 978-1-119-11073-6 (ebk)

Manufactured in the United States of America

10 9 8 7 6 5 4 3 2 1

For general information on our other products and services please contact our Customer Care Department within the United States at (877) 762-2974, outside the United States at (317) 572-3993 or fax (317) 572-4002.

Wiley publishes in a variety of print and electronic formats and by print-on-demand. Some material included with standard print versions of this book may not be included in e-books or in print-on-demand. If this book refers to media such as a CD or DVD that is not included in the version you purchased, you may download this material at http://booksupport.wiley.com. For more information about Wiley products, visit www.wiley.com.

Library of Congress Control Number: 2016947909

To Nilay, Burak, Semra, and Mustafa Yener, for all your support and the time I needed to write this book.

—MURAT

To Canan and my entire family: Aysel, Ismail, Ugur, Umut, Aysun, and Murat.

—ONUR

ABOUT THE AUTHORS

MURAT YENER is a code geek, open source committer, Java Champion, and Google Developer Expert on Android, who is working at Intel as an Android developer. He is the author of *Professional Java EE Design Patterns* (Wrox, 2015). He has extensive experience with developing Android, Java, web, Java EE, and OSGi applications, in addition to teaching courses and mentoring. Murat is an Eclipse committer and one of the initial committers of the Eclipse Libra project.

Murat has been a user group leader at GDG Istanbul since 2009, organizing, participating, and speaking at events. He is also a regular speaker at major conferences such as DroidCon, JavaOne, and Devoxx.

LinkedIn: www.linkedin.com/in/muratyener

Twitter: @yenerm

Blog: www.devchronicles.com

ONUR DUNDAR started his professional career in software engineering at Intel Corporation as a Software Application Engineer working on mobile platforms, tablets, and phones with MeeGo and Android, and later on IoT platforms such as IPTV, Intel's Galileo, and Edison. He is the author of *Home Automation with Intel Galileo* (Packt Publishing, 2015). Onur presents training sessions on Android application development, so he was keen to author this book.

LinkedIn: www.linkedin.com/in/odundar

Twitter: @odunculuk

Amazon author page: www.amazon.com/Onur-Dundar/e/B00V0VOIGA

ABOUT THE TECHNICAL EDITOR

XAVIER HALLADE is Application Engineer at Intel Software and Services Group in France. He's been working on a wide range of Android frameworks, libraries, and applications. Xavier is a Google Developer Expert in Android, with a focus on the Android NDK, and actively contributes in the Android community, writing articles, helping developers on Stack Overflow, and giving talks around the word.

CREDITS

PROJECT EDITOR
Tom Dinse

TECHNICAL EDITOR
Xavier Hallade

PRODUCTION EDITOR
Barath Kumar Rajasekaran

COPY EDITOR
Nancy Rapoport

MANAGER OF CONTENT DEVELOPMENT & ASSEMBLY
Mary Beth Wakefield

PRODUCTION MANAGER
Kathleen Wisor

MARKETING MANAGER
Carrie Sherrill

PROFESSIONAL TECHNOLOGY & STRATEGY DIRECTOR
Barry Pruett

BUSINESS MANAGER
Amy Knies

EXECUTIVE EDITOR
Jody Lefevere

PROJECT COORDINATOR, COVER
Brent Savage

PROOFREADER
Nancy Bell

INDEXER
Johnna VanHoose

COVER DESIGNER
Wiley

COVER IMAGE
Leo Blanchette/Shutterstock

ACKNOWLEDGMENTS

I WANT TO THANK MY COAUTHOR, ONUR DUNDAR, for all his hard work and for his efforts to keep me on schedule. Without him, this book wouldn't be half as good.

I am grateful to my team and colleagues, in particular John Wei and Sunil Tiptur Nataraj, who gave me the time and flexibility to work on this book, and Angus Yeung for his support. My thanks also go to my Google Developer Relations contacts, Uttam Tripathi, Martin Omander, Baris Yesugey, and others who supported me in all ways possible. I want to thank Alex Theedom for covering for me on my Java EE–related responsibilities while I was deeply buried in Android. No words are enough to thank Jim Minatel and Tom Dinse, who patiently worked on all the details while keeping most of the stresses away from us. And thanks, of course, to everyone at Wrox/Wiley who got this book on the shelves.

I must thank three important people who are responsible for who I am in my professional life.

First, thanks to my dad, Mustafa Yener, for giving me my first computer, a C64, at an early age while I was asking for slot cars. I wrote my very first code on that computer.

Second, thanks to my thesis advisor, Professor Mahir Vardar, for the early guidance I needed to start my career.

Finally, thanks to my lifetime mentor and friend, Naci Dai, who taught me almost everything I know about being a professional software developer.

—MURAT YENER

I WANT TO THANK MURAT YENER for taking me on this journey and enabling me to share in the fun of writing this book. With Murat's experience and knowledge, we delivered a high quality book for developers and engineers.

I am thankful to Oktay Ozgun for his wisdom and advice throughout my career and life. His guidance has helped me become a passionate engineer and a better person.

I also want to thank Professor Arda Yurdakul, who was generous with her experience and guidance as she encouraged me to learn more about software and computers and to become a better engineer.

In my final year at Bogazici University, Professor Cem Ersoy, Professor Alper Sen, and Dr. Hande Alemdar provided great assistance. I would also like to thank the folks at Intel—Steve Cutler, Andrew John, Brendan LeFoll, Todor Minchev, Peter Rohr, Rami Radi, Alex Klimovitski, and Marcel Wagner—for their great support and for the opportunities they provided to help me learn and develop in my engineering career.

I want to thank Professor Gurkan Kok, for the interesting stuff I am working on right now.

Thanks to all the Wiley/Wrox editors for their help with this book.

Finally, my special thanks to Ali Caglar Ozkan for motivating me (and taking great photos) after my first book, along with with Aytac Yurdakurban and Engin Efecik, for supporting my continuing interest in writing books.

—ONUR DUNDAR

CONTENTS

FOREWORD

"A bad workman always blames his tools."

When my brother and I were growing up in leafy, suburban England, my father relied on that old proverb. My brother and I often found excuses to avoid the admittedly small number of chores we were asked to do, and we placed the blame for not completing the task at hand on an inanimate object that couldn't answer back. This proverb was my father's standard response to our excuses, and it immediately negated our protestations.

As I've gotten older, I have learned that there is far more wisdom in this old proverb than merely getting young boys to complete their chores. It is not just about ensuring that you have the right tools for the task at hand, but it is also about having the knowledge of how to use them effectively that is key to being productive. If you think about the standard tools in a handyman's toolkit, it is pretty clear to most people that a hammer is not the most appropriate tool to remove a screw from a piece of wood, but as tools become more complex and refined, these distinctions become less clear.

In the Android development world, the de-facto standard development tool is Android Studio, not least because it is under extremely active development by Google—the same organization responsible for Android itself. Since the early preview versions arrived in May 2013, the feature set has grown quite considerably and continues to do so at an impressive rate. If we also consider that Android Studio is built on the foundations of IntelliJ IDEA—which is already an extremely feature-rich development environment—then it should be pretty clear that any analogies with hammers or screwdrivers are going to break down rather quickly. Rather than comparing Android Studio with individual tools, it is, perhaps, better to consider it as the entire toolbox, which contains lots of individual tools that can sometimes be used individually, sometimes be used together, but, when used effectively, can simplify and speed up many of our everyday development tasks—including the really mundane or repetitive ones that we all hate!

Modern software development is so much more than simply writing code, and this is especially true on Android. The main logic of your app may be written in Java. You also have resources (which are largely XML-based) such as vector drawables (which incorporate SVG path data into that XML), build files (which are groovy/grade files), and test source code (which is Java with test domain–specific dialects such as Espresso, Fest, or Hamcrest). This is before you start considering frameworks that change the syntax and flow of your code, such as Rx, and even alternate languages that are gaining traction, such as Kotlin. Mastery of all of this can be hard. The ability to "context switch" between different components, languages, frameworks, and dialects is made much easier by basic features such as code highlighting and pre-compilation, which show errors inline as you code. But we are so used to these that we hardly notice them, and because they have become second nature to us, context switching itself becomes second nature.

While using the tools available until they become second nature is important, a prerequisite for that is actually knowing what tools there are and how to use them effectively. That is where this book comes in. Murat and Onur have provided a guide to Android Studio and its many facets that will be of great value to both the novice and the seasoned Android Studio user alike.

MARK ALLISON
June 2016

INTRODUCTION

NO MATTER HOW GOOD YOU ARE AT WRITING CODE, without proper knowledge of Integrated Development Environments (IDEs), you will face many obstacles. This book covers Google's Android Studio, the official tool for developing Android applications. Each chapter focuses on a specific topic, progressing from the basics of how to use the IDE to advanced topics such as writing plugins to extend the capabilities of the IDE.

WHO THIS BOOK IS FOR

This book is for developers with any level of experience. Whether you are new to Android or a seasoned Android developer who used Eclipse-based ADT before, this book will bring you to a level where you can unleash your true development potential by making use of Android Studio's tools.

WHAT THIS BOOK COVERS

This book not only covers features of Android Studio essential for developing Android apps but also touches on topics related to the whole development cycle. The following are just a few examples of the topics covered that are basic to Android Studio or that extend its capabilities:

➤ Sharing and versioning your code with Git

➤ Managing your builds with Gradle

➤ Keeping your code maintainable and bug free with testing

➤ Controlling the whole build and test cycle with Continuous Integration

➤ Writing plugins for Android Studio to extend its capabilities and add desired custom features

➤ Using third-party tools with Android Studio to improve the development process

HOW THIS BOOK IS STRUCTURED

Each chapter focuses on a specific topic related to Android Studio or an accompanying tool by explaining why it is needed and how it is used or configured. Some chapters provide code samples to demonstrate the use case or provide an example for the topic.

➤ **Chapter 1: Getting Started:** Installing and setting up your development environment. Creating an emulator for running your projects.

➤ **Chapter 2: Android Studio Basics:** Beginning with Android Studio, creating a new project, building your project, and migrating projects to Android Studio.

➤ **Chapter 3: Android Application Development with Android Studio:** Structure of Android Studio projects. How to use assets, XML files and the Android Manifest. Creating and working with modules.

➤ **Chapter 4: Android Studio In Depth:** Deep dive into Android Studio, explaining menus, editors, views, and shortcuts. How to use live templates and refactoring. How to build your projects and sign apks.

➤ **Chapter 5: Layouts with Android Studio:** How to use layouts with Android Studio. Explanation of previews and tools for UI development. Managing external dependencies. How to use and organize assets.

➤ **Chapter 6: Android Build System:** How to use and configure Gradle effectively. Writing plugins for Gradle

➤ **Chapter 7: Multi-Module Projects:** Adding modules in your project. How to create and work with Phone/Tablet, Library, Wear, TV, Glass, Auto, and Cloud modules.

➤ **Chapter 8: Debugging and Testing:** Debugging Android code with ADB. Learn details of the Android Devices Monitor, Android virtual devices, Lint, and testing your code.

➤ **Chapter 9: Using Source Control: GIT:** How to share your project and enable version control by using Git.

➤ **Chapter 10: Continuous Integration:** Automating your builds, tests, and releases using continuous integration servers.

➤ **Chapter 11: Using Android NDK with Android Studio:** Installing and using Android NDK for building projects with C/C++ code.

➤ **Chapter 12: Writing Your Own Plugins:** Writing your own plugins to extend the capabilities of the IntelliJ platform. Interacting with UI, editor, and adding your actions.

➤ **Chapter 13: Third-Party Tools:** Other accompanying tools that can help and speed the development lifecycle.

WHAT YOU NEED TO USE THIS BOOK

Any modern computer with an operating system that is supported by Android SDK and Android Studio is sufficient to use Android Studio, build Android apps, and run the samples given in this book. You need to install appropriate Android SDK, Android Studio, and Java Virtual Machine (JVM) for your OS. Some chapters require additional tools or frameworks to be installed such as Android NDK. You can find more information on exact hardware requirements needed in Chapter 1.

WHY WE WROTE THIS BOOK

In November 2007, Google released a preview version of Android SDK to allow developers to start playing with the new mobile operating system. Roughly two years later, in October 2009, ADT (Android Developer Tools) a plugin set for Eclipse, was released to the public.

As a Google I/O 2009 attendee, I (Murat) was lucky enough to have an Android device and was probably one of the earliest developers to download and install the plugins to my Eclipse. As years passed, we both followed the same passion to download and try new stuff released with new ADT versions.

At the time, I was an Eclipse committer who knew how to write plugins, extend the IDE's capabilities, and introduce the behavior and functionality I needed. So with each release of ADT, I was more and more excited to see what had been done with the tools.

On May 2013, at Google I/O, roughly four years after our love-hate relationship with ADT started, Google announced Android Studio, which soon became the official, supported IDE for Android development. ADT was never perfect. but it was familiar. Like many other developers, we knew all the shortcuts, how things work, what to do when something was not working, workarounds, and how the projects were structured. More significantly, we were able to write our own plugins or inspect ADT plugins to see why something went wrong. However, with the release of Android Studio, suddenly we were all in a new platform that we knew very little about.

We resisted switching to Android Studio for a while, but finally gave it a try. Suddenly, Android, a platform we were long familiar with, was a stranger. The new project structure was very different because of the changes introduced by IntelliJ and Gradle. To adopt IntelliJ, we decided to follow IntelliJ shortcuts instead of using IntelliJ shortcut mapping for Eclipse shortcuts, which made the situation even worse. We were barely able to search for a file or piece of code, navigate through menus, right-click to create files, or even generate some basic getters and setters. We went from being experts with ADT to beginners with Android Studio.

We had finally had enough! We were experienced developers, but struggled with Android Studio and were not able to show our skills. So we started following IntelliJ talks, pinning the IntelliJ shortcut cheat sheet in our cubicles, reading IntelliJ plugin code, and forcing ourselves to use Android Studio in our daily work.

This book is the summary of the lessons we learned walking unaided on this difficult path . This book is what we needed for ourselves when we were switching from Eclipse-based ADT to IntelliJ-based Android Studio. This is why we believe any developer, whether an Android newbie or a seasoned Android developer who used to work on ADT, will find this book useful for developing his or her knowledge of the tools that are actually there to support his or her coding skills.

Quoting Alex Theedom, co-author of my previous book: "Every chapter that we wrote has this goal: Write content that we would like to read ourselves." We followed the same goal with Onur and the result is the book you are holding in your hands.

We hope that you enjoy reading this book as much as we enjoyed writing it.

> **NOTE** *Be sure to read our blog at* `http://www.devchronicles.com/2016/06/` `expert-android-studio-book-updates.html` *to see the changes announced at Google I/O 2016.*

CONVENTIONS

To help you get the most from the text and keep track of what's happening, we've used a number of conventions throughout the book.

> **WARNING** *Boxes like this one hold important, not-to-be forgotten information that is directly relevant to the surrounding text.*

> **NOTE** *Notes, tips, hints, tricks, and asides to the current discussion are offset and placed in italics like this.*

As for styles in the text:

➤ We *highlight* new terms and important words when we introduce them.

➤ We show keyboard strokes like this: Ctrl+A.

➤ We show file names, URLs, and code within the text like so: `persistence.properties`.

➤ For code:

```
We use a monofont type for code examples.
We use bold to emphasize code that is of particular importance in
the current context.
```

SOURCE CODE

As you work through the examples in this book, you may choose either to type in all the code manually or to use the source code files that accompany the book. All of the source code used in this book is available for download at www.wiley.com/go/expertandroid. Once at the site, simply click the Download Code link on the book's detail page to obtain all the source code for the book.

> **NOTE** *Because many books have similar titles, you may find it easiest to search by ISBN; this book's ISBN is 978-1-119-08925-4.*

Once you download the code, just decompress it with your favorite compression tool. Alternately, you can go to the main Wrox code download page at http://www.wrox.com/dynamic/books/download.aspx to see the code available for this book and all other Wrox books.

ERRATA

We make every effort to ensure that there are no errors in the text or in the code. However, no one is perfect, and mistakes do occur. If you find an error in one of our books, such as a spelling mistake or a faulty piece of code, we would be very grateful for your feedback. By sending in errata, you may save another reader hours of frustration and at the same time you will be helping us provide even higher quality information.

To find the errata page for this book, go to http://www.wrox.com and locate the title using the Search box or one of the title lists. Then, on the book details page, click the Book Errata link. On this page you can view all errata that have been submitted for this book and posted by Wrox editors. A complete book list including links to each book's errata is also available at www.wrox.com/misc-pages/booklist.shtml.

If you don't spot "your" error on the Book Errata page, go to www.wrox.com/contact/techsupport.shtml and complete the form there to send us the error you have found. We'll check the information and, if appropriate, post a message to the book's errata page and fix the problem in subsequent editions of the book.

P2P.WROX.COM

For author and peer discussion, join the P2P forums at p2p.wrox.com. The forums are a Web-based system for you to post messages relating to Wrox books and related technologies and interact with other readers and technology users. The forums offer a subscription feature to e-mail you topics of interest of your choosing when new posts are made to the forums. Wrox authors, editors, other industry experts, and your fellow readers are present on these forums.

At `http://p2p.wrox.com` you will find a number of different forums that will help you not only as you read this book, but also as you develop your own applications. To join the forums, just follow these steps:

1. Go to `p2p.wrox.com` and click the Register link.

2. Read the terms of use and click Agree.

3. Complete the required information to join as well as any optional information you wish to provide and click Submit.

4. You will receive an e-mail with information describing how to verify your account and complete the joining process.

> **NOTE** *You can read messages in the forums without joining P2P but in order to post your own messages, you must join.*

Once you join, you can post new messages and respond to messages other users post. You can read messages at any time on the Web. If you would like to have new messages from a particular forum e-mailed to you, click the Subscribe to this Forum icon by the forum name in the forum listing.

For more information about how to use the Wrox P2P, be sure to read the P2P FAQs for answers to questions about how the forum software works as well as many common questions specific to P2P and Wrox books. To read the FAQs, click the FAQ link on any P2P page.

1

Getting Started

WHAT'S IN THIS CHAPTER?

➤ System requirements for Android Studio

➤ Java installation instructions for Microsoft Windows, Mac OSX, and Linux

➤ Android Studio installation instructions for Microsoft Windows, Mac OSX, and Linux

In this chapter you get started with setting up your development environment so you can start Android development with Android Studio. To that end, this chapter covers the basic installation instructions for Android Studio and its system requirements.

Although the Android operating system is based on Linux, the Android SDK and tools are available for all major operating systems, so you can set up your development environment for the operating system you are working with. Throughout this book we use Mac OS as the main environment; however, we cover Linux and Windows setup as well.

SYSTEM REQUIREMENTS FOR WINDOWS, MAC OS X, AND LINUX

To use Android Studio, your development system must meet the minimum system requirements. This section lists the minimum requirements for Windows, Mac OS X, and Linux.

Microsoft Windows

➤ Microsoft Windows 10/8/7/Vista/2003 (32 or 64 bit)

➤ 2GB RAM minimum, 4GB RAM recommended

- ➤ 400MB hard disk space
- ➤ At least 1GB for Android SDK, emulator system images, and caches
- ➤ 1280 × 800 minimum screen resolution
- ➤ Java Runtime Environment (JRE) 6 or higher
- ➤ Java Development Kit (JDK) 7
- ➤ Optional for accelerated emulator: Intel processor with support for Intel VT-x, Intel EM64T (Intel 64), and Execute Disable (XD) Bit functionality

Mac OS X

- ➤ Mac OS X 10.8.5 or higher, up to 10.9 (Maverick)
- ➤ 2GB RAM minimum, 4GB RAM recommended
- ➤ 400MB hard disk space
- ➤ At least 1GB for Android SDK, emulator system images, and caches
- ➤ 1280 × 800 minimum screen resolution
- ➤ Java Runtime Environment (JRE) 6
- ➤ Java Development Kit (JDK) 7
- ➤ Optional for accelerated emulator: Intel processor with support for Intel VT-x, Intel EM64T (Intel 64), and Execute Disable (XD) Bit functionality

Linux

- ➤ GNOME or KDE desktop
- ➤ GNU C Library (glibc) 2.15 or later
- ➤ 2GB RAM minimum, 4GB RAM recommended
- ➤ 400MB hard disk space
- ➤ At least 1GB for Android SDK, emulator system images, and caches
- ➤ 1280 × 800 minimum screen resolution
- ➤ Java Runtime Environment (JRE) 6 or higher
- ➤ Oracle Java Development Kit (JDK) 7

More details about system requirements can be found at `https://developer.android.com/sdk/index.html#Requirements`.

Keep in mind that based on the size of the project, number of your dependencies, and emulator usage, you will likely need more resources. Typically, you will need at least 8GB of RAM and GPU support to run an emulator and work smoothly with better compilation times.

In most cases, developers need to test applications on multiple devices. Because they usually don't have enough devices to test adequately, they rely on emulators. Emulators require a high amount of storage and memory to run faster.

> **NOTE** *Emulators are virtual devices, so having a CPU with virtualization support is crucial for developers to get the best experience with Android emulators.*

INSTALLING JAVA

Java is essential for all operating systems. You must install Java SE (Standard Edition) Development Kit (JDK) for your operating system.

> **NOTE** *We suggest the Java distribution provided by Oracle. It is possible to encounter problems with OpenJDK or other Java distributions.*

> **NOTE** *At the time of this writing, Java SE 8 is the latest version of JDK.*

The JDK 7 download page can be accessed directly at `http://www.oracle.com/technetwork/java/javase/downloads/jdk7-downloads-1880260.html`. When you navigate there, the page shown in Figure 1-1 appears.

To download the required installation binary or packages, select the Accept License Agreement option and then click the download link of the binary or package for your operating system.

The following sections provide installation instructions for Oracle Java version 7 on 64 bit Windows, Mac OS X, and Linux.

Installing Java for Windows OS

Java installation on Windows is pretty straightforward. As mentioned in the previous section, installing JDK provides JRE as well.

> **NOTE** *Depending on the version of your Windows installation, you need to download and install either the 64 bit or 32 bit version of the JDK. Since most modern computers are equipped with 64 bit CPUs, we will continue with installation of the 64 bit version. If you have a 32 bit Windows installation, use the 32 bit JDK, which is listed as the Windows x86 version.*

Download the jdk-7u79-windows-x64.exe file and run it to start the installation.

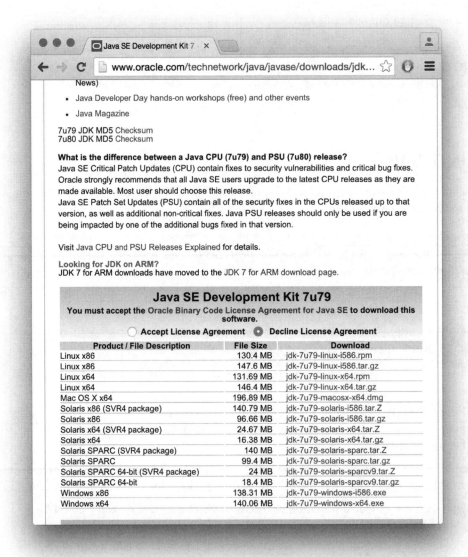

FIGURE 1-1: JDK download page

Install Java

Start the Java installation by clicking the jdk-7u79-windows-x64.exe file after downloading it; you will see the dialog box shown in Figure 1-2.

FIGURE 1-2: Java installation setup wizard

1. Click the Next button to continue.

2. In the window that opens, all items are selected by default to be installed on your local hard drive. You can change the installation path and which installation modules to install. For Android application development, the Public JRE and Development Tools options must be selected if they are not already installed on your machine. (If they are already installed, they will not be listed inside the window.) You may deselect Source Code, which is used to install public Java API classes. It is not mandatory to install the source code. Make your selections in the dialog shown in Figure 1-3.

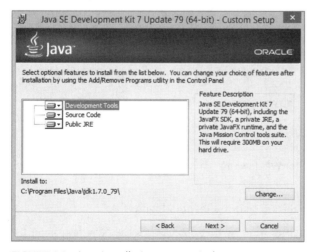

FIGURE 1-3: Java installation setup window

3.   We suggest that you continue with the default selections and use the default installation path, then click Next to finish the installation.

Setting Paths for Java on Windows 10

After the installation finishes, you need to set the Windows 10 environment paths for Java to config-ure the system environment. You need to set a path for the JAVA_HOME system variable.

> **WARNING** *In earlier versions of Windows, the steps for setting the environ-ment path might be a little different.*

1.   Right-click the Start menu icon and click File Explorer. In the window that opens, right-click This PC and select Properties to open the System window.

2.   From the options at the left of the System window, select Advanced system settings. This will open the dialog box shown in Figure 1-4 with the Advanced tab enabled.

FIGURE 1-4: Java path setup for Windows 10

3.   Click the Environment Variables... button shown in Figure 1-4.

4.   From the Environment Variables window that opens, click the New button and set the Java path with your installation directory. As shown in Figure 1-5, the path is C:\Program Files\Java\jdk1.7.0_79 for our 64 bit installation. If you installed the x86 version, your path would be different, such as C:\Program Files (x86)\Java\jdk1.7.0_79.

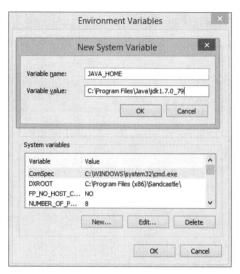

FIGURE 1-5: Java Environment Variables setup for Windows 10 64-bit

Installing Java for MacOS X

Java used to be a part of Mac OS X and was shipped by Apple. This changed several years ago. Apple also decided to remove Java from Mac OS so JDK, which is provided by Oracle, needs to be installed separately.

1. Start by downloading the `jdk-7u79-macosx-x64.dmg` file from the page shown in Figure 1-1.

2. Launch the dmg file to display the window shown in Figure 1-6.

FIGURE 1-6: Mac OS X Java installation

3. Double click the JDK 7 Update 79.pkg file link to start the installation.

4. Select the installation directory you want and complete the installation in the window shown in Figure 1-7.

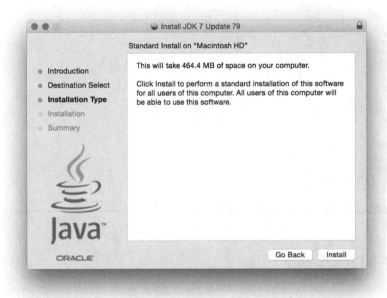

FIGURE 1-7: Java installation Max OS X

Now JDK is ready to use on Mac OS X. You can check the installed JDK version from System Preferences of Mac OS X or type `'java -version'` in the terminal window to see whether Java installed properly.

Installing Java for Linux

Two types of installation packages are available for Linux. If you use a distribution with an rpm package manager like Fedora, you can download the rpm package and install Java via rpm. In this section you install JDK with the `tar.gz` package on Ubuntu 14.04.

1. Download `jdk-7u79-x64.tar.gz` from the download page shown in Figure 1-1 and extract it to the directory where you downloaded the `tar.gz` file using the following command:

   ```
   user@ubuntu$ tar xzvf jdk-7uXX-x64.tar.gz
   ```

   That command will extract JDK into the `jdk1.7.0_79` folder where you ran the command.

2. Move that folder to `/usr/local/java` with the following command:

   ```
   user@ubuntu$ mv jdk1.7.0_79 /usr/local/java
   ```

3. Edit the `/etc/profile` file to set Java paths for your session. You can copy and paste the following lines to the end of the `/etc/profile` file.

```
##Java Path
JAVA_HOME=/usr/local/java/jdk1.7.0_79
JRE_HOME=$JAVA_HOME/jre
PATH=$PATH:$JAVA_HOME/bin:$JRE_HOME/bin
export JAVA_HOME
export JRE_HOME
export PATH
```

4. Install Java binaries for system-wide use with the following commands:

   ➤ Install the Java binary:

```
user@ubuntu$ sudo update-alternatives --install "/usr/bin/java" "java"
"/usr/local/java/jdk1.7.0_79/bin/java" 1
```

   ➤ Install the Java Compiler binary `javac`:

```
user@ubuntu$ sudo update-alternatives --install "/usr/bin/javac" "javac"
"/usr/local/java/jdk1.7.0_79/bin/javac" 1
```

   ➤ Install the Java Web Start binary `javaws`:

```
user@ubuntu$ sudo update-alternatives --install "/usr/bin/javaws" "javaws"
"/usr/local/java/jdk1.7.0_79/bin/javaws" 1
```

5. Set Oracle Java as the default Java for your system with the following commands:

```
user@ubuntu$ sudo update-alternatives --set java
/usr/local/java/jdk1.7.0_79/bin/java

user@ubuntu$ sudo update-alternatives --set javac
/usr/local/java/jdk1.7.0_79/bin/javac

user@ubuntu$ sudo update-alternatives --set javaws
/usr/local/java/jdk1.7.0_79/bin/javaws
```

When you are done with the previous instructions, JDK and JRE will be ready to use when you restart Ubuntu. You can test whether Java installed correctly with version control. The command and output for that will look like this:

```
user@ubuntu~$ java -version
java version "1.7.0_79"
Java(TM) SE Runtime Environment (build 1.7.0_79-b15)
Java HotSpot(TM) 64-Bit Server VM (build 24.79-b02, mixed mode)
```

INSTALLING ANDROID STUDIO

Android Studio installation, like Java installation, differs by operating system. The following sections provide installation instructions for Windows, Mac OS X, and Linux platforms.

The direct link for the installation binaries is `https://developer.android.com/sdk/index.html`.

The download link that's available when you go to this site will be correct for the operating system you are running, as shown in Figure 1-8.

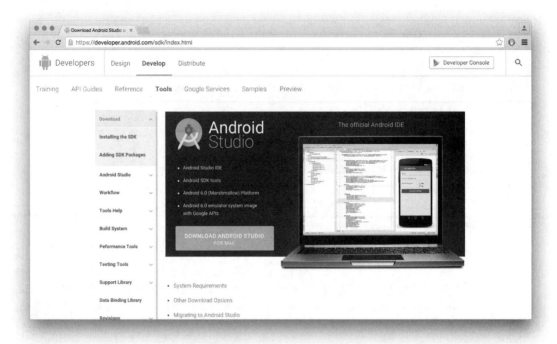

FIGURE 1-8: Android Studio download page

> **NOTE** *Download options have installers bundled with Android SDK tools.*

Installing Android Studio for Microsoft Windows 10

This section covers the installation instructions for Android Studio on Microsoft Windows 10.

> **WARNING** *Make sure you are connected to Internet while you are installing the Android Studio because installation process includes downloading required Android SDK installation files from the web.*

1. Click the Download Android Studio for Windows link to download the latest Android Studio installer exe file.

2. Run the exe file after the download completes. You will see the window shown in Figure 1-9.

FIGURE 1-9: Android Studio Setup window

3. Click the Next button to select installation components. The Android Studio option can't be changed, but you can deselect the Android SDK, Emulator, and Intel HAXM installations, as shown in Figure 1-10.

FIGURE 1-10: Android Studio Setup configuration for Windows

4. The next window prompts you for installation paths for Android Studio and Android SDK separately.

5. As shown in Figure 1-11, the installation asks about Intel HAXM memory configuration, and lets you choose a custom memory configuration.

FIGURE 1-11: Intel HAXM configuration dialog for Windows

> **TIP** *Recommended memory for HAXM is 2GB, but you can change that based on your hardware. We recommend that you install HAXM if you plan to use the emulator.*
>
> *HAXM is a hardware-assisted virtualization engine that lets you use your computer's processor to generate x86 Android images. Without HAXM, the emulator's performance will greatly suffer.*

You are now ready to launch Android Studio on Windows. The first time you launch Android Studio, it asks you to select the theme for the IDE, as shown in Figure 1-12.

Finally, the installation completes and Android Studio is ready to work on Android application projects.

Installing Android Studio for Mac OS X

This section covers the basic steps to install Android Studio on Mac OS X.

1. From the page shown in Figure 1-8, click the Download Android Studio for MAC link.

   Download the `android-studio-ide-141.2178183-mac.dmg` file, which includes the Android Studio IDE installer for Mac OS.

2. Launch the file you just downloaded.

3. Drag and drop the Android Studio.app icon into Applications folder, as shown in Figure 1-13.

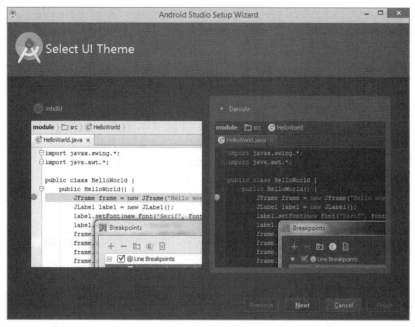

FIGURE 1-12: Android Studio theme selection on Windows

FIGURE 1-13: Android Studio installer for Mac OS X

After copying Android Studio to the Applications folder, Android Studio is ready to launch. You can then remove the `.dmg` file from your system.

When you first launch Android Studio, it asks if you want to install Android SDK, the Android emulator, and Intel HAXM. It will also ask for the setup path for them, as shown in Figure 1-14.

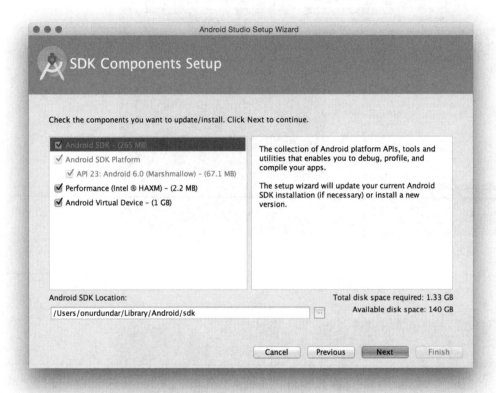

FIGURE 1-14: Android Studio Setup Wizard for Max OS X

If you selected Intel HAXM installation, you are asked for the amount of RAM memory you want to make available for the virtual devices, as shown in Figure 1-15.

After you click Finish from the Emulator Settings dialog box, a window with an installation summary will display as shown in Figure 1-16.

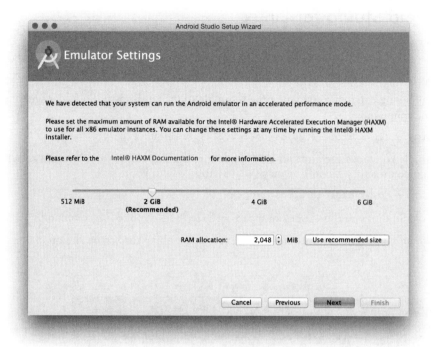

FIGURE 1-15: Intel HAXM configuration for Mac OS X

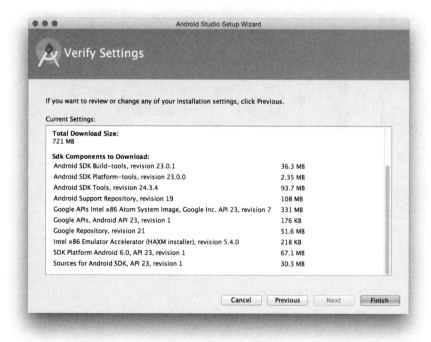

FIGURE 1-16: Summary window for Max OS X installation

Installing Android Studio for Linux

Navigate to the download page shown in Figure 1-9 to download Android Studio for Linux (`android-studio-ide-141.2178183-linux.zip`). After you've downloaded the file, follow these steps:

1. Enter the following command to extract the setup file to the android-studio folder where you executed the command:

    ```
    user@ubuntu$ unzip android-studio-ide-141.2178183-linux.zip
    ```

    In this example you move the android-studio folder to the `/opt` directory. You can select your own home directory as well, to make it available only to you.

    ```
    user@ubuntu$ sudo mv android-studio /opt
    ```

2. Start Android Studio with the `./opt/android-studio/bin/studio.sh` command.

    When you first launch Android Studio on Linux, it will display the screen shown in Figure 1-17.

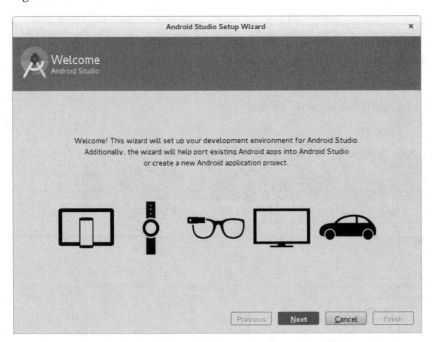

FIGURE 1-17: Android Studio Setup Wizard for Linux

Then the setup wizard will ask for Standard or Custom installation.

3. Select Custom installation to see the installation packages.

    The wizard moves to the window shown in Figure 1-18 where you can select an Android Studio UI theme.

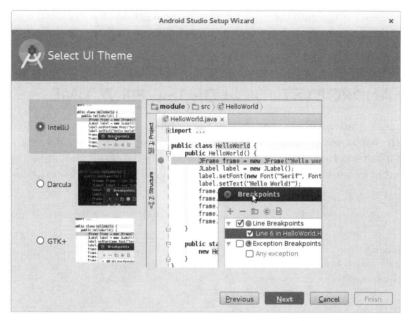

FIGURE 1-18: Theme selection window for Linux

Figure 1-18 shows that the IntelliJ theme has been selected for this installation.

4. Select the Android Studio, Android SDK, and Emulator as shown in Figure 1-19.

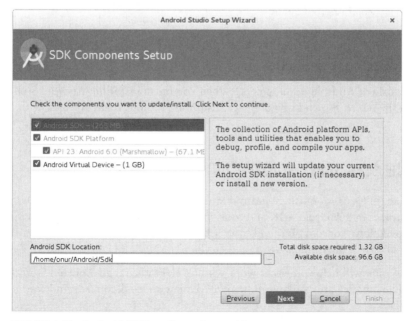

FIGURE 1-19: Android SDK configuration on Linux

5. Indicate the installation path for Android SDK in the Android SDK Location field at the bottom of the dialog box, as shown in Figure 1-19. Click Finish to complete the Android SDK installation.

LAUNCHING ANDROID STUDIO FOR THE FIRST TIME

When you first launch Android Studio, you will see the Complete Installation dialog box shown in Figure 1-20, which enables you to import settings from a previous installation. Because we made a clean installation for this example, we selected the last option in the dialog box. If you have a previous installation with customization you'd like to import, you can specify your previous installation path (see the first two options in Figure 1-20).

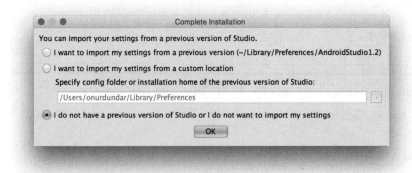

FIGURE 1-20: First launch of Android Studio

Welcome to Android Studio

Welcome to the world of Android development! When you finish installing Android Studio, you will finally reach the screen shown in Figure 1-21. Android Studio is ready to work with Android projects.

STANDALONE SDK INSTALLATION

In this book, the main focus is on using Android Studio for development at an advanced level. Therefore, the book covers Android SDK and tools installations together with Android Studio. If you would like to explore Android SDK and tools separately, you can get the standalone installation binaries for your choice of operating system.

Standalone installation will help you either work with an IDE other than Android Studio, or to use the tools alone. The binaries are available at http://developer.android.com/sdk/index.html at the bottom of the page.

FIGURE 1-21: Welcome to Android Studio

As you did in this chapter, you can download the compatible binary for your operating system and follow similar installation steps to continue. After you download and extract the SDK, you can add its location to Android Studio using Settings ⇨ Appearance & Behavior ⇨ System Settings ⇨ Android SDK ⇨ Android SDK Location.

SUMMARY

In this chapter, we wanted to make sure you have all the necessary tools to work on the examples in the following chapters. We started by providing the requirements for the basic computer system needed to install the required software.

We then covered the installation of Android Studio for Windows, Mac OS, and Linux. We continue with what is required to begin Android application development in Android Studio in Chapter 2.

2

Android Studio Basics

WHAT'S IN THIS CHAPTER?

➤ How to create an Android Studio project

➤ Android project structure

➤ Creating and configuring virtual devices

➤ Building and running your project via ADB

➤ Migrating Android projects from Eclipse IDE

Welcome to Android Studio! Whether you are an Eclipse-based ADT veteran or a total newbie to Android development, you will enjoy Android Studio, which is based on IntelliJ IDEA and offers new tools, a UI editor, a whole new build system, memory/CPU analyzers, and many more new features and functionalities.

After long years of the Eclipse-based ADT plugin suite, Google announced (at Google I/O 2014) that Android Studio would be the official supported IDE for Android Development. Of course, you can still use ADT if you are coming from a strong Eclipse background; however, you will probably face problems that you may need to solve on your own.

As an Eclipse committer who has written code for several Eclipse projects, I preferred to stay with ADT for a long time. However, with the announcement that Android Studio is the official IDE, "resistance is futile."

If you are new to Android development, it may even be easier to adapt to Android Studio because your previous Eclipse experience might not necessarily help you a lot on the IntelliJ platform.

CREATING A NEW SAMPLE PROJECT

Figure 2-1 shows the welcome screen of Android Studio, which is the first screen you will see when you launch Android Studio. From here, you can start a new project. In this chapter, you will work on a sample project, so let's open an existing project by importing the code.

FIGURE 2-1: Welcome to Android Studio window

1. Click the Import an Android code sample option in the list shown in Figure 2-1.

   Numerous sample projects are hosted in GitHub. Because those projects are always up-to-date and new ones are added when a new API or functionality is introduced, it is a good idea to use them as a reference template for your own projects. Plus, all those samples are under the Apache software license, which makes them available to be used freely even in commercial products.

Because everyone loves selfies, in this example you build yet another selfie app.

2. Scroll down to Camera and select Camera2 Basic from the list shown in Figure 2-2.

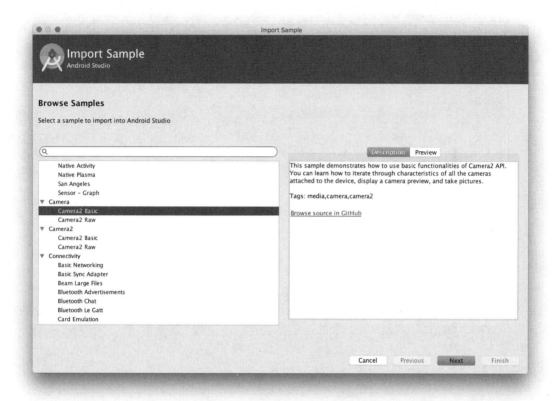

FIGURE 2-2: Import Sample—Browse Samples window

Keep the Application name and Project location that appear on the Sample Setup screen (see Figure 2-3).

3. Click Finish.

Your project will be downloaded from GitHub, and the main IDE window, which is mostly empty, will appear.

Once you click Finish, Android Studio creates the project and switches to the main development layout. Before you figure out where to find your project files, let's take a look at Android SDK configuration in the following sidebar.

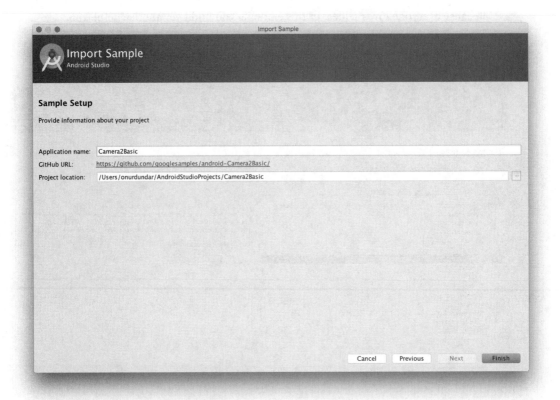

FIGURE 2-3: Import Sample—Sample Setup window

ANDROID SDK CONFIGURATION INSIDE ANDROID STUDIO

Android SDK can be downloaded and configured within Android Studio too with using Android Studio Preferences window. The following steps walk you through the configuration process.

1. Click the SDK Manager icon on the right side of the top toolbar, as shown in Figure 2-4.

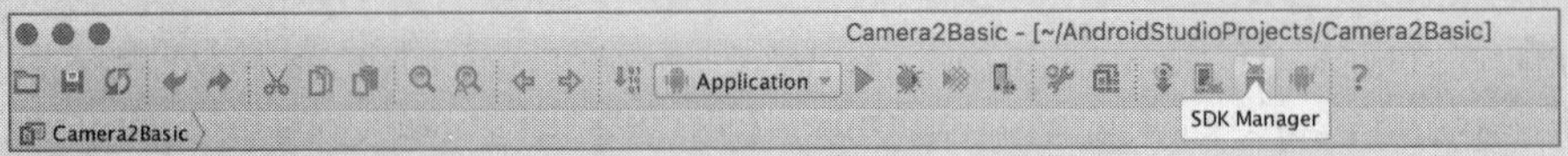

FIGURE 2-4: SDK Manager button in Android Studio

The Android Studio Default Preferences window opens, as shown in Figure 2-5. The window focuses on the SDK integration option, listing the installed and available SDK versions as well as showing if any of them are eligible for an update.

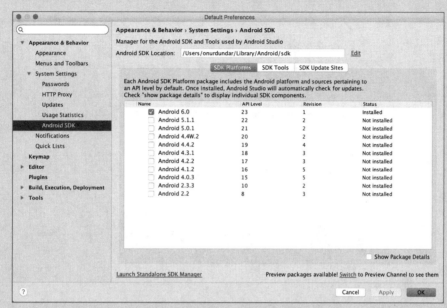

FIGURE 2-5: Android Default Preferences for Android SDK window

2. You can use the checkboxes next to installed versions and click Apply to start the installation of the desired version. As shown in Figure 2-5, a typical setup may include installed, not installed, and partially installed SDK versions.

You can also choose to start the standalone SDK Manager by clicking the Launch Standalone SDK Manager link shown at the bottom of Figure 2-5. After clicking that link, you see the detailed installation options shown in Figure 2-6.

continues

continued

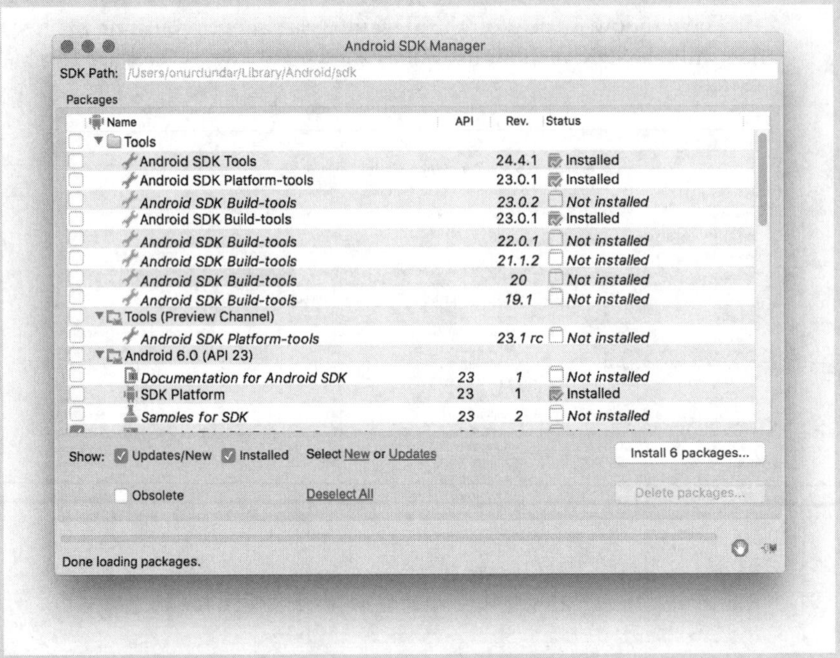

FIGURE 2-6: Android SDK Manager installation window

NOTE *We recommend that you install and update to the latest version of Android SDK Tools and Android SDK Platform tools. The SDK version you should use mostly depends on your project requirements. As a starting point, we recommend installing the latest SDK, which is API 23, and a widely accepted version such as API 19 or 20.*

3. Scroll down to Extras and make sure Support Repository, Support Library, Google Play Services, Google Repository, Google USB Driver (only for Windows), and Intel x86 Emulator Accelerator are selected.

4. Once you make sure all components are selected, click the Install button and accept the license. The SDK Manager will start the download, which may take a while depending on your selections.

NOTE *Android SDK Manager can also be used for deleting unused SDKs and build tools.*

Alternatively, you can use the Android Studio Preferences view and select the SDK version you want to install.

Using Different SDKs

When the SDK installations are complete, you can start working with one of them. As long as your Android project does not make use of a feature that is introduced with a specific version of an SDK, you can easily convert your project to work with an older SDK.

> **NOTE** *We recommend that you work with the latest version of the SDK to build and compile your project, but use the* minSDK *attribute to support the earliest supported version.*

> **NOTE** *We cover build and support versions of the SDK in this and following chapters.*

Android Project Structure

Everything looks great? Well, not exactly because you should probably be looking at an empty screen, as shown in Figure 2-7.

> **NOTE** *We say that "you should probably be looking at an empty screen" because we want these instructions to remain version agnostic. It is possible that future updates might cause a change and your screen might not be empty.*

Although the initial project screen shows nothing about your project, Android Studio gives you a list of hints about how to move to the next step. For this example, press Command+1 on Mac or Alt+1 on Windows to open the project view. The project view, shown in Figure 2-8, displays all the contents of your project.

> **WARNING** *If you are coming from an ADT background, be aware that Android project structure has changed dramatically with Gradle. With Android Studio, project resources are grouped by types, which does not correspond to their locations on the file system. Although this is a clever approach and is handy, it can also be tricky if you are used to the projects view from ADT.*

If you prefer to list resources similar to the way they are hosted in the file system, click the Android list on the top left and a menu with different options will open. Choose Project, and Android Studio will group your project resources as they appear in the file system, as shown in Figure 2-9.

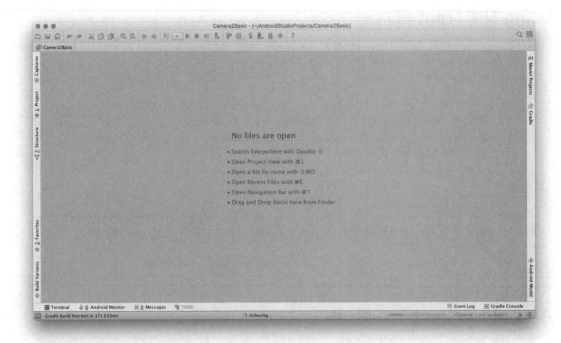

FIGURE 2-7: Android Studio after importing the sample project

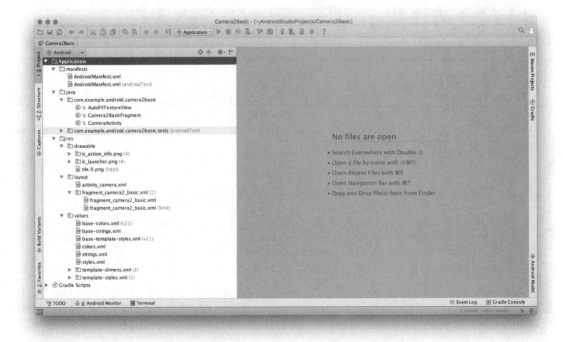

FIGURE 2-8: Project view on Android Studio

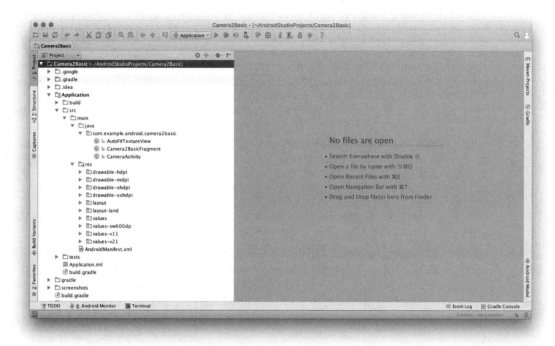

FIGURE 2-9: Traditional project view

Let's examine this view to gain a better understanding of Android Studio project structure. Every project has a few hidden folders, which you might not be able to navigate with your file manager.

Typically, an Android project has three hidden folders, as listed at the top left of Figure 2-9. The hidden folders are:

➤ **.idea folder**—This folder keeps IntelliJ-specific project metadata and settings not necessarily shared with source control systems (so not shared to someone else).

➤ **.gradle folder**—This folder keeps Gradle-related bin files. This folder's contents are not subject to change unless you change the project's Gradle version.

➤ **.google folder**—This folder includes sample packing files from Google.

> **NOTE** *Typically, users should not directly edit any file in the hidden folders. Let the IDE deal with them.*

Next is the application folder. The name of this folder can change depending on your application name and preferences. However, you can easily recognize the folder because it has a small device symbol on the lower right of the folder icon. This type of folder holds the source code, application files, and configuration.

There can be more than one Application folder in your project depending on its size and architecture. In this chapter we will assume there's one named "app," but will dig into different combinations in later chapters.

Expand the app folder by clicking the triangle to the left of the folder name. If you have developed Android applications before, the contents should be familiar to you. If not, the following pages give detailed information.

The project folders are:

➤ **build folder**—This folder might be the least important of all because, as a developer, you won't need to deal with or edit anything inside it. The Gradle build system will be triggered to build your source files by the IDE and produce the output into this folder.

➤ **src/main folder**—This folder might be the most important folder because it hosts all your source code except for tests. If you expand the src folder, you may see package folders that group your source files. We will explain this later in this chapter.

➤ **src/test and src/androidTest folders**—These folders might be the most underestimated in the whole Android project. The basic convention with tests is to place Unit tests into src/test and instrumentation tests into src/androidTest folders. They hold your test files, which can be run during compilation, packaging, or even on a build server. Good test coverage for your source files is needed if you want to keep your code maintainable, open to change, and still bug free!

There are also several files in the root of the project folder. These are essential because they usually affect each module in the project. You may need to edit the following essential project files.

➤ **build.gradle**—Although each module in a project has its own build.gradle file, the top level build.gradle is inherited by each of them. Any global Gradle setting for a repository or a library can be added to this file.

➤ **local.properties**—Each user has an SDK and NDK file path in their computer. For example, say you work for a corporation where you need to have proxy settings, including your credentials. Adding that personal data to a Gradle file, which would be added to source control, may not be wise. Such info can be added to local.properties and kept out of source control.

➤ **settings.gradle**—Most Android projects have multiple modules, which may consist of libs or wear extensions. Once a build is executed, Gradle checks settings.gradle to figure out which projects need to be included in the build.

You may find additional files in the root project folder, which you don't need to edit or worry about for now. Although we covered all root level files and folders, we haven't covered the most important one, the src folder.

The src folder hosts all source, resource, and application manifests. Expand the source folder to list its contents, as shown in Figure 2-10.

Inside the src folder is only one folder, which is named main. The main folder contains the java and res folders, which have different icons than other folders to highlight their importance.

The java folder contains all the packages in the format of reverse URL and Java classes. In our example, we have only one package, com.example.android.camera2basic, which has three Java classes. Clicking a class file will open the editor and display the chosen Java file's contents, as shown in Figure 2-11.

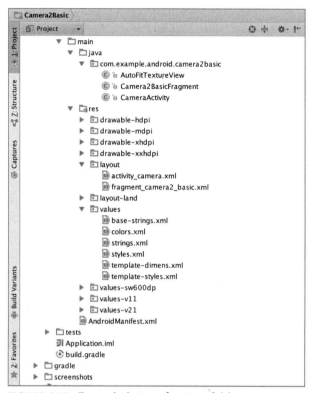

FIGURE 2-10: Expanded view of project folders

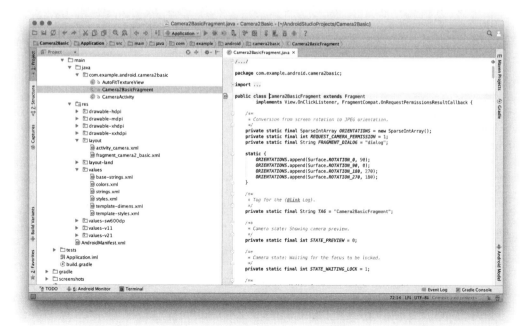

FIGURE 2-11: Opened Java file on Android Studio

We cover the editor in detail but first let's move to the other folder inside main. The res folder holds all resource files, including images, layouts, localization files, and so on. Android projects have different folders for different screen sizes, pixel densities, and other parameters, as shown in Figure 2-12.

FIGURE 2-12: res folder in Android project view

Placing different sizes of the same image into a drawable folder will leverage the ability of Android to display the most appropriate image for the device your application is running on. The idea is the same for the layout and values folders. Different layouts can be added for landscape and portrait views, and different values can be added for different versions and pixel densities.

This approach has given Android the capability to run on different screen sizes and densities from the beginning, unlike most other mobile platforms, which used to offer only a fixed resolution.

> **TIP** *When you are developing for Android always keep in mind that your application may target very different screen sizes—from phones to tablets, as well as watches, TVs, and even glasses.*

Finally, we can focus on one final file that is very trivial for an Android application, the `AndroidManifest.xml` file. The Android manifest holds metadata from the list of activities, services, application name and version, target and minimum SDK requirements, and hardware requirements for the target devices, as well as the permissions that your app requires. Listing 2-1 shows the contents of `AndroidManifest.xml`.

LISTING 2-1: AndroidManifest.xml Content

```xml
<?xml version="1.0" encoding="UTF-8"?>
<!--
Copyright 2014 The Android Open Source Project

Licensed under the Apache License, Version 2.0 (the "License");
you may not use this file except in compliance with the License.
You may obtain a copy of the License at
```

```
          http://www.apache.org/licenses/LICENSE-2.0

     Unless required by applicable law or agreed to in writing, software
     distributed under the License is distributed on an "AS IS" BASIS,
     WITHOUT WARRANTIES OR CONDITIONS OF ANY KIND, either express or implied.
     See the License for the specific language governing permissions and
     limitations under the License.
     -->

  <manifest xmlns:android="http://schemas.android.com/apk/res/android"
        package="com.example.android.camera2basic">

     <uses-permission android:name="android.permission.CAMERA" />

     <uses-feature android:name="android.hardware.camera" />
     <uses-feature android:name="android.hardware.camera.autofocus" />

     <application android:allowBackup="true"
          android:label="@string/app_name"
          android:icon="@drawable/ic_launcher"
          android:theme="@style/MaterialTheme">

        <activity android:name=".CameraActivity"
                 android:label="@string/app_name">
           <intent-filter>
              <action android:name="android.intent.action.MAIN" />
              <category android:name="android.intent.category.LAUNCHER" />
           </intent-filter>
        </activity>
     </application>

  </manifest>
```

Our manifest file starts with the manifest declaration, which also declares the main package. This declaration enables us to refer to subpackages and classes by using only the suffix after the root package.

Next, the manifest declares the permissions, followed by the uses-feature tag to declare the hardware requirements of the application.

Every activity and service component that resides in an Android project must be listed under the application tag. The sample project consists of only one activity, which is used as the entry point of the sample app, so the activity is listed as .CameraActivity, only with the full path and name after the root package and with the LAUNCHER intent. This activity will be used for launching the application presented in the Android manifest.

Building and Running a Project

The sample project is a complete and ready-to-run application, so we can move on to building and running the application. Android Studio offers different ways to compile and run projects. To simply build a project, select Build from the toolbar and then the Make Project option, as shown in Figure 2-13.

FIGURE 2-13: Build menu list

Although this option will compile and package your app, it will not execute your app on either a device or the emulator. To run the sample project, select Run from the toolbar and then the Run Application option, as shown in Figure 2-14.

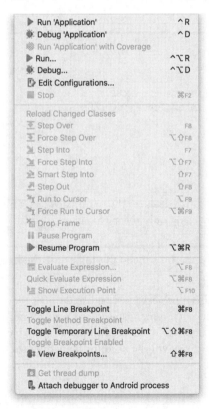

FIGURE 2-14: Run menu items in Android Studio

Alternatively, you can click the green Play icon (or Control+R on Mac, Ctrl+R in Windows), as shown in Figure 2-15.

FIGURE 2-15: Android Studio Run 'Application' button

Depending on your computer's hardware, Android Studio will spend some time to build the application and later will ask for a target device, which can also be an emulator, to run the application on. You haven't either created a virtual device or connected a real device yet, so you'll do that in the next section.

ANDROID EMULATOR

Android Emulator is a great tool that is bundled with Android Studio. It enables your computer to emulate Android hardware and operating system to run your apps and provide a preview of how it would behave on a real device. The Android emulator enables you to test your application on a variety of screen sizes, hardware configurations, Android versions, and even different CPU architectures.

However, the Android emulator has a bad reputation for being very slow. Developers used to make fun of it by saying, "If you optimize your app for the emulator, it will run smoothly on any device." This may sound exaggerated but was almost true in the past.

The main performance problem behind the emulation resulted from the ARM CPU emulation on personal computers, which mostly run on x86 CPU architecture. In 2011, Intel introduced HAXM and Google started providing x86-based Android system images, which boosted the emulator performance by using the host CPU instead of emulating a different CPU architecture.

Installing HAXM

To start using the emulator, you need to install HAXM and the x86 image of the desired version of Android. HAXM requires a minimum version of Android SDK 17. In most cases, HAXM installation is pretty straightforward. If you already installed the SDK (see Chapter 1), you can start creating your virtual device. If you didn't install it, you can set it up with the SDK Manager, as discussed in the "Android SDK Configuration Inside Android Studio" sidebar earlier in this chapter.

1. Click Launch Standalone SDK Manager (refer to Figure 2-5) to open the Standalone SDK Manager shown in Figure 2-6.

2. Scroll down to Extras and select Intel x86 HAXM Emulator Accelerator (HAXM Installer), as in Figure 2-16.

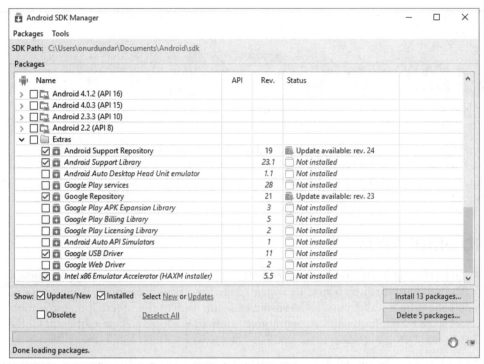

FIGURE 2-16: Intel HAXM selection in Windows

3. Once the download of HAXM is complete, you need to manually trigger its installer, from `sdk\extras\intel\Hardware_Accelerated_Execution_Manager\` on Windows, or `sdk/extras/intel/Hardware_Accelerated_Execution_Manager/` on Mac.

Creating a New Android Virtual Device

It is very easy to create a new Android device. AVD (Android Virtual Device) Manager is the next icon. It's to the left of the SDK Manager, as shown in Figure 2-17.

FIGURE 2-17: Android Studio AVD Manager button

1. Click the AVD Manager button. Because you haven't created a virtual device before, AVD Manager is currently empty (see Figure 2-18).

2. Click the Create Virtual Device button, shown in Figure 2-18, to start creating a virtual Android device.

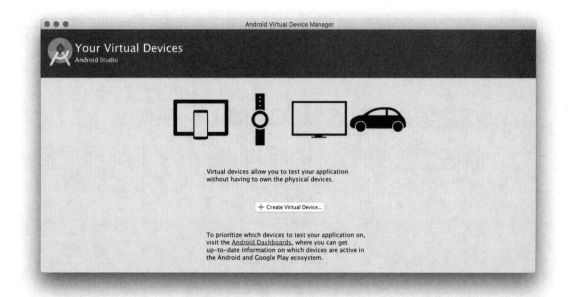

FIGURE 2-18: AVD Manager's initial appearance

A list of available devices appears, as shown in Figure 2-19. At the top of the list are Nexus devices, which are the reference devices released by Google. The rest of the list contains common screen sizes and device properties. You are free to modify any device from the list or even create your own for testing purposes. Creating your own device might be a good idea for testing devices that you don't have access to. In addition to creating virtual devices to simulate phones, AVD Manager also supports tablets, wear, and TV.

3. Select Nexus 5X. The Nexus 5x is one of the two reference devices released with Android 6.0. Although you will continue with Nexus 5X here, you can choose any device to create a virtual device.

4. Select Marshmallow and make sure the target column (see Figure 2-20) includes "(with Google APIs)." In this step, you are free to choose either the ARM or Intel-based Android images listed in Figure 2-20.

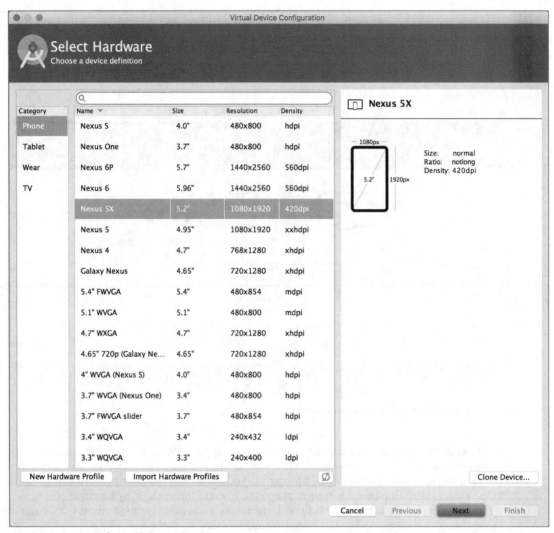

FIGURE 2-19: List of virtual device hardware

Nexus 5X is an ARM-based device. Although selecting ARM will provide more accurate device/virtual device testing, emulating ARM on an x86-based laptop will require additional memory and processing power and will result in performance issues. By installing HAXM, you can have your virtual device run an Intel image to provide better performance.

5. Click the Next button shown in Figure 2-20 to tweak final settings of your virtual device.

6. Make the final configurations, as shown in Figure 2-21, and click Finish.

That is it—you created a virtual device that can run your sample project. To run the emulator, open AVD Manager and click the Play icon shown in the Actions column of the virtual device, as shown in Figure 2-22.

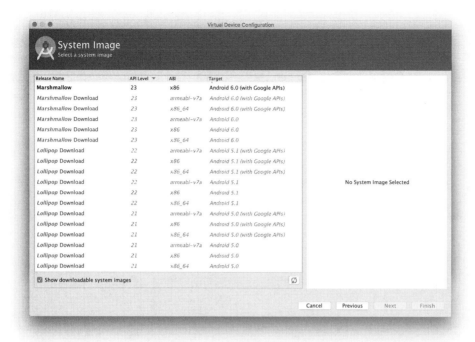

FIGURE 2-20: System Image selection for AVD

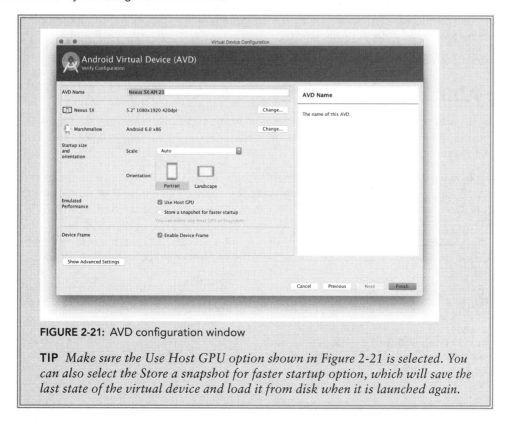

FIGURE 2-21: AVD configuration window

TIP *Make sure the Use Host GPU option shown in Figure 2-21 is selected. You can also select the Store a snapshot for faster startup option, which will save the last state of the virtual device and load it from disk when it is launched again.*

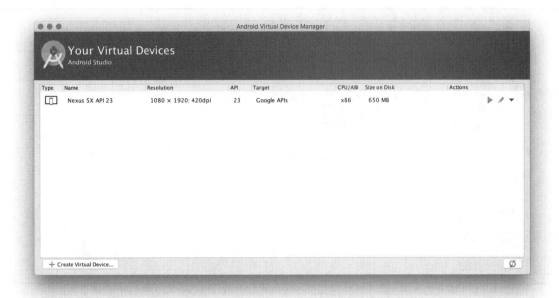

FIGURE 2-22: List of created virtual devices

Depending on your hardware configuration, the emulator may take a while to boot up. The initial bootup will take longer than subsequent launches, which might be close to instantaneous if you have chosen the Store a snapshot for faster startup option.

You now have a new Android device to play with and it didn't cost you a penny.

Using ADB

You have now imported and built your project and created a new Android virtual device. In this section, you learn how to deploy the binary to the virtual device. To do that, you need to know how to connect the two.

Luckily, you won't need to do much; Android Debug Bridge (ADB) will handle all the heavy lifting here. ADB works almost seamlessly to deploy and run your app in a virtual or real device. However, you can always access ADB through your OS's shell and execute commands manually. ADB bridges between your computer and the connected device whether it is a real or a virtual device. Most of the time, you won't need to interact with ADB manually. To use ADB, open a console and type **adb**. You should see something similar to Listing 2-2 followed by the usage and option details.

LISTING 2-2: ADB LAUNCH COMMAND

```
$ adb
Android Debug Bridge version 1.0.32
Revision eac51f2bb6a8-android
...
```

> **NOTE** *If the* adb *command isn't found, you'll have to add it to your system's path. The* adb *executable is located in the* sdk/platform-tools *folder.*

Given that the emulator you just created is still running, if you execute adb devices you'll see a list of the devices available over ADB. You can use adb install <apk path> to install a packaged apk to the connected device. However, because the IDE performs this for you, you won't be using the install option extensively.

Other useful adb options are push and pull. These commands are used to access the device's file system. The following commands will copy a file from your computer to the device and from your device to the computer:

```
$ adb push <local> <remote>
$ adb pull <remote> <local>
```

MIGRATING PROJECTS FROM ECLIPSE

Eclipse ADT and Android Studio have very different project structures and configurations, but importing projects from Eclipse to Android Studio is very straightforward in most cases.

The first option for migrating your Eclipse project to IntelliJ is to import the project into Android Studio.

1. Select File ➪ New ➪ Import Project as shown in Figure 2-23.

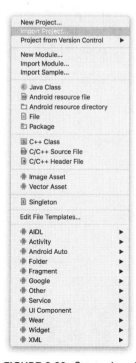

FIGURE 2-23: Start migrating from Eclipse

2. Navigate through your Eclipse project folder and click OK, as shown in Figure 2-24. The IDE will create the necessary files, including Gradle files, and set up your project.

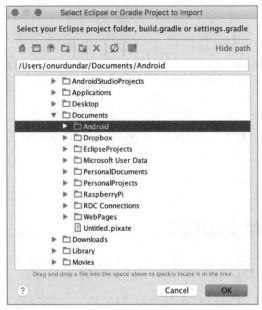

FIGURE 2-24: Select project path to import

Once the import is complete, `import-summary.txt` will be shown, which displays the results of the migration.

Another way to migrate your project is to export the project from Eclipse. To export from Eclipse, your ADT plugins should be up-to-date.

1. Right-click on your project and select Export, as shown in Figure 2-25.

FIGURE 2-25: Export from Eclipse

2. Navigate to the Android folder group and select Generate Gradle build files, as shown in Figure 2-26.

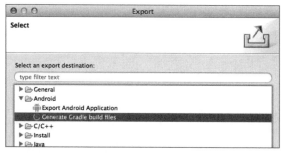

FIGURE 2-26: Export window on Eclipse

3. Follow the instructions in the wizard that opens. That will generate the Gradle files needed for Android Studio.

Although your project is converted to Gradle by the end of this process, the project still needs to be imported to Android Studio for IntelliJ-specific files to be created.

Both methods work on most of Android projects without any problem. However, if you happen to come across a problem in either method, switch to the other method to complete the migration.

SUMMARY

This chapter gave you a basic overview of the whole Android Studio project structure, Gradle, and ADB. We have seen how to create and configure virtual Android devices using an emulator and how to run your applications using ADB.

We also covered how to migrate Eclipse ADT-based Android projects to new Android Studio–based Gradle projects.

3

Android Application Development With Android Studio

WHAT'S IN THIS CHAPTER?

- ➤ Android projects
- ➤ Android activities
- ➤ Android services
- ➤ Assets of Android projects
- ➤ XML files in Android projects
- ➤ Android manifest file
- ➤ Creating and working with modules
- ➤ Building modules for Android Projects

Android Studio is your best friend when it comes to Android application development and distribution. Android Studio doesn't only help you writing code but it also provides tools and templates to fasten your development process. Android Studio is increasingly more capable than a traditional Android development environment in Eclipse with the Android Developer Tools (ADK) plugin.

This chapter guides you through Android application development in Android Studio. We cover the basic building blocks of Android applications and the capabilities of Android Studio.

We start by creating an Android application project for phones and tablets and continue with additional development modules for your application.

ANDROID PROJECTS

Software applications with many source files and resources are organized under a project structure or a folder structure to better classify files and define source code compiling and binary generation order with a hierarchy. We call these predefined folder and file structures a *project*, which is an organizational unit that represents a complete software solution.

Android applications also have a file and folder hierarchy structure, referred to as an Android application project. Android application projects include Java source code, XML configuration files, images, videos, sounds, and other resources within an organizational structure. For each unique application you develop, you need to create a new project. Android Studio helps from the first step when creating the initial project structure.

The build process uses the configuration files and folder hierarchy to identify Java source and resource file relations in the Android Project to create the final application package to run on an Android device. If you don't follow the correct structure, you won't be able to create the final application package and your application simply won't build.

In Chapter 2, you saw how to import an example project. In this chapter, you create a new empty project with auto-generated files for Android Studio's application build process.

Creating a New Android Project

Creating a new Android project is pretty straightforward with Android Studio. If you have just launched the Android Studio, it will display the screen shown in Figure 3-1 and list the Quick Start options.

> **NOTE** *If Android Studio is already running, you need to select File ⇨ New ⇨ New Project from the Android Studio menu to create a new Android Studio project.*

1. To begin, click Start a new Android Studio project.

2. Name your application. Your application's name will be shown in the list of applications on the Android menu to launch the application. It's conventional to start an application name with a capital letter. As you can see in Figure 3-2, we named this example application ChapterThree.

> **NOTE** *If you are going to distribute your application in the Play Store, the application's name should be unique, catchy, and easy to search for.*

FIGURE 3-1: Android Studio Welcome window

3. Configure your domain name for packaging application source files. Reserved domain name syntax is used by Android Studio to create a package name and store Java source files.

Sun, later acquired by Oracle, recommends that developers use a company domain written in reverse for the package name to prevent name collisions for the Java classes. By default, Android Studio shows com.example, which is overwritten after you enter the domain name. You don't need to buy a domain name for your personal project but you need to make sure the domain name you are about to choose is unique.

We named our first example's company domain with our book's title expertandroid.com (refer to Figure 3-2). You may choose your company's name as the domain name or just write any name.

4. Accept the default project location and click Next to select the device type and SDK version you want to deploy your application to. Deployment targets can be phones, tablets, wearables, TVs, or Android Auto applications. Throughout the book, Android SDK 6.0 (Marshmallow) is used, so for this example we selected Android SDK 6.0 for Phone and Tablet, as shown in Figure 3-3.

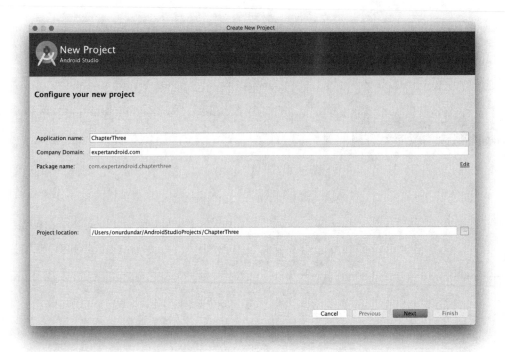

FIGURE 3-2: Create New Project window

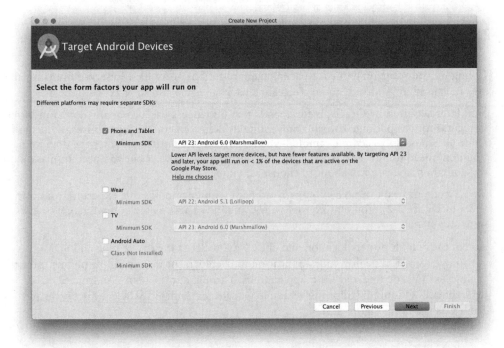

FIGURE 3-3: Target device selection window

You can select previous versions of Android SDK according to your application's requirements. It is not always the best option to select the latest version for Android applications because not all the devices on the market are updated for the latest version. Earlier SDK versions are compatible with the latest releases so if you select an earlier version, your application will work on the latest devices.

Android Studio can help you select the best SDK version for your application. Click the Help me choose link shown in Figure 3-3 to see the Android Platform/API Distribution. The window shown in Figure 3-4 will appear.

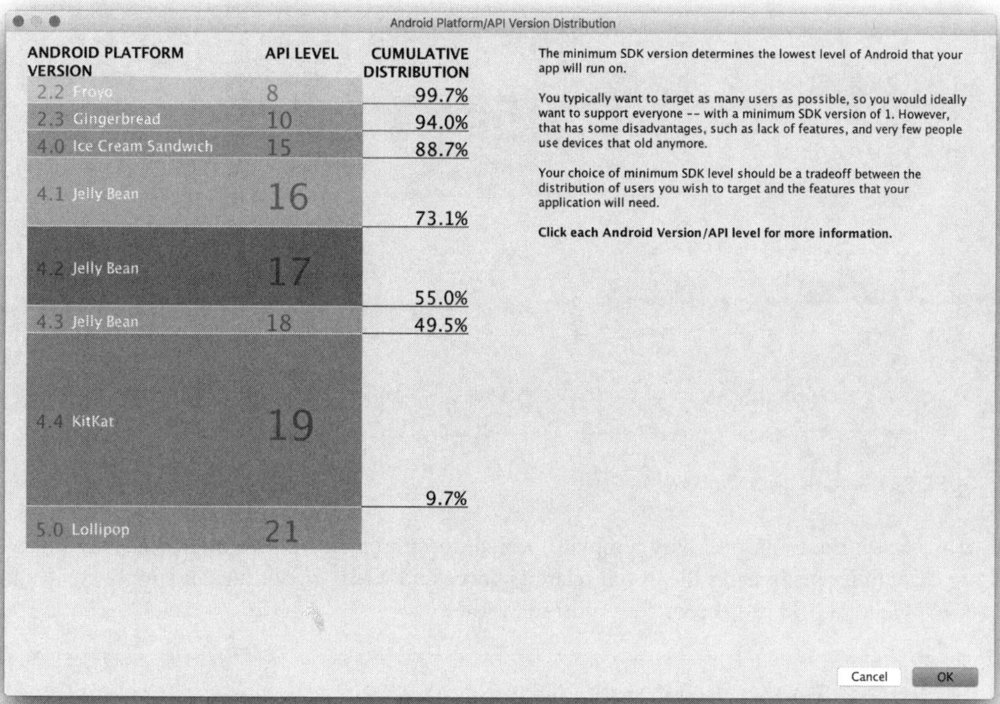

FIGURE 3-4: Android Platform/API Version Distribution window

The distribution list might help you to decide about your application's target audience, but remember that the latest devices from Google always have the most recent Android distribution and get updated regularly.

Also, be aware of device proliferation. There are thousands of different Android devices and most have different specifications, so you need to test your application for several of them.

5. After selecting the device and SDK version, click OK to display the window shown in Figure 3-5. Here you select the initial activity for the application.

FIGURE 3-5: Add activity window

6.　Select the Empty Activity template, which doesn't include a user interface. The Empty Activity just creates the initial related source and XML layout files. It simply creates the Hello World project for the Android Studio.

> **NOTE** *You can also select Add No Activity and choose to add activities at a later stage in the development process. The Add No Activity option is also used when you want to create a service application without a user interface or a library, which would be consumed by other projects with activities.*

7.　After you select the Empty Activity option, click Next to open the window shown in Figure 3-6. This is where you name your activity. Each activity selection requires different naming and configuration options. The Empty Activity only needs naming and layout generation selection. We selected to generate the main activity layout, which is the user interface configuration file, so we named our first activity MainActivity.

　　All related resources will use the activity's name so, for example, the layout xml file and other configuration files will include "Main" in their names.

After you click Finish, Android Studio generates files and folders for the application and configures the application build environment. When the project is complete, it is displayed in Android Studio, as shown in Figure 3-7.

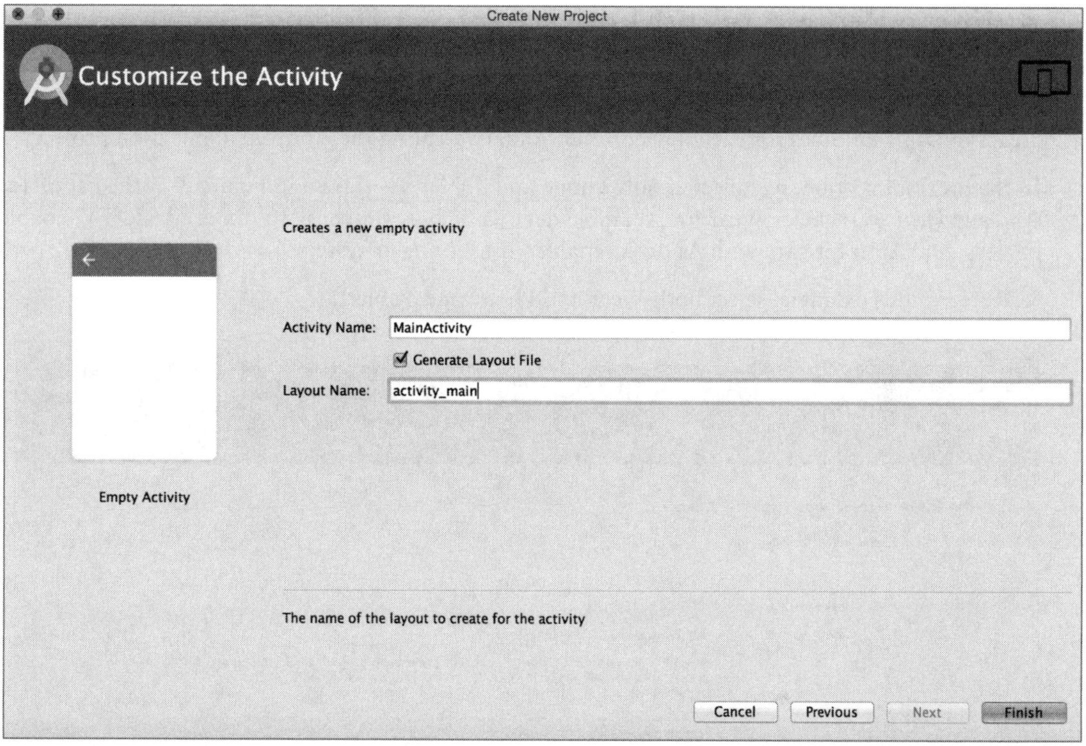

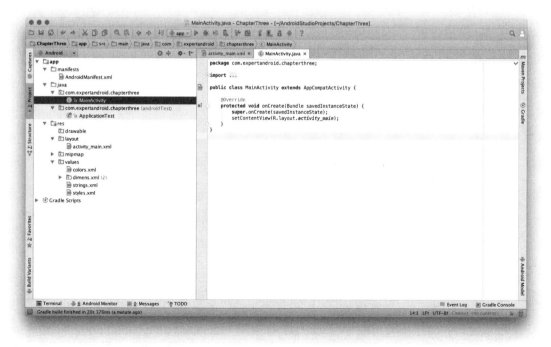

FIGURE 3-6: Activity customization window

FIGURE 3-7: First Project View in Android Studio

Creating a Project with Multiple Target Devices

In the past few years, the Android operating system has expanded to devices such as set top boxes, TVs, wearable devices such as watches and glasses, and even automobiles. Android Studio helps you create multiple modules for each device while you are creating the Android application project.

In the previous section, we selected only Phone and Tablet. As shown in Figure 3-8, the list of Target Android Devices includes Wear for wearable devices such as watches, TV for Android TV–enabled devices, and Auto for cars with Android-enabled infotainment systems.

1.   For this example, select both Wear and Phone and Tablet.

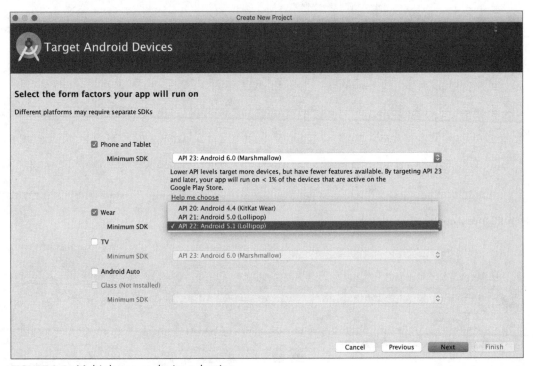

FIGURE 3-8: Multiple target device selection

2.   Just like applications for Phone and Tablet, applications for other devices should have an activity. The setup wizard for wearable devices offers a variety of activities for you to choose from, as shown in Figure 3-9.

3.   The setup wizard prompts you to name the activity and populate the rest of the fields for the project. Note that Wear Activity has more fields to initialize. Name the activity MainActivity, as shown in Figure 3-10.

> **NOTE** *Using the same name for both the Wear Activity and the Phone and Tablet application activity doesn't create a conflict because they are handled in different directories.*

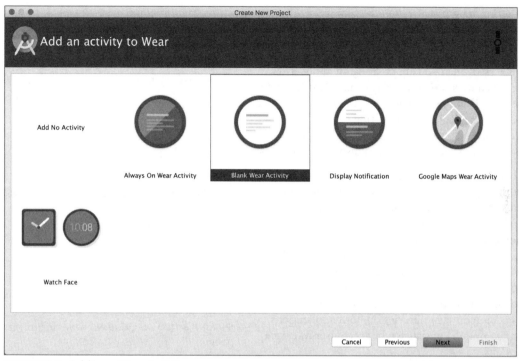

FIGURE 3-9: Adding a Wear Activity

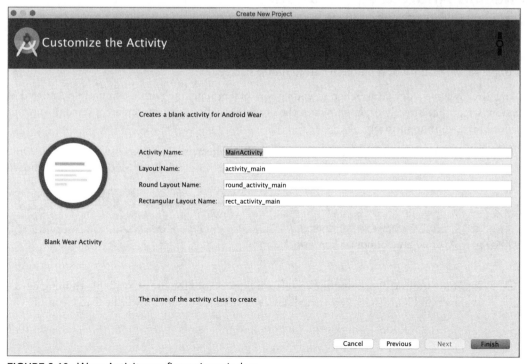

FIGURE 3-10: Wear Activity configuration window

4.   Click Finish and after the project is initialized, you will see two modules in Android Studio, as shown in Figure 3-11.

Having two target platforms in one project is suggested when you are creating an application designed to run on both platforms. Health and fitness applications are a good example of these kinds of projects: A fitness application on a phone or tablet receives a user's running and walking data from the Android watch device and displays the details.

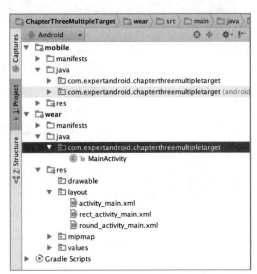

To accomplish this cooperation between devices, both target platforms' source and resource files must be in the same project, but they will be divided into different modules. Android application architecture requires Wear application binaries to be distributed inside the phone/tablet binary package, which is facilitated when you have both apps in a single project. This structure will make you design better communication between devices, use common Java libraries on both applications, and maintain consistency between user interface designs.

FIGURE 3-11: Two modules shown in Project View

Launching Android Applications

After you have created the application project, you need to build the application and launch it on a device. (You learn about configuring it in the following sections of this chapter.) The build process is covered in Chapter 2; here we cover launching your first application with Android Studio. This example shows Hello World on the Android phone's screen.

If you already have a device attached to your development machine with a compatible Android SDK version, select that directly or, as shown in the previous section, you can create a virtual Android device to run your application.

1.   To run an Android application from the initial project, click the Run 'app' arrow shown in Figure 3-12. Alternatively, you can type Shift+F10 (Control+R on Mac) on your keyboard.

FIGURE 3-12: Android Studio toolbar Run 'app' button

2.   Android Studio will prompt you to select a device from the list of available running devices, as shown in Figure 3-13. If a virtual device is already running, it will be listed as well.

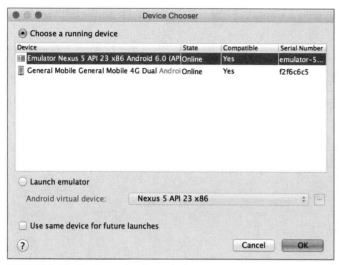

FIGURE 3-13: Device selection window

If an emulator is not running, you can select the Launch emulator option and click OK to launch the virtual device and run the application, as shown in Figure 3-14. If you intend to use the same device again for running or debugging your project, you can check the Use same device for future launches box.

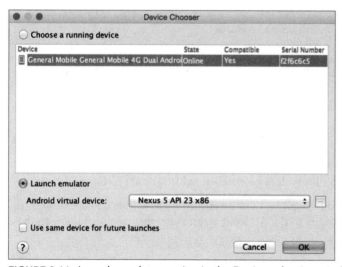

FIGURE 3-14: Launch emulator option in the Device selection window

The application will automatically launch and you will see the activity you created running on the device.

If you followed this example, you will see an empty application printing "Hello World!" to the screen, as shown in Figure 3-15.

FIGURE 3-15: Hello World application running on the emulator

The following section covers the details of other activity templates that can be added to Android projects.

ANDROID ACTIVITIES

Android is a Linux-based operating system; all applications running on the Android devices are Linux processes and each application's lifecycle is predetermined from start to finish.

Activities are the most fundamental building blocks of an Android application. There is no `main` function defined for Android applications so the Android system launches applications with the defined launcher activity, which serves as the entry point for the application.

When the application is launched, the Android operating system application manager uses the Main Activity to start the application. The Main Activity is the first activity users see.

Android activities have a defined lifecycle to manage application runtime from launch to the end of application life. Figure 3-16 shows the simple states of an activity's lifecycle.

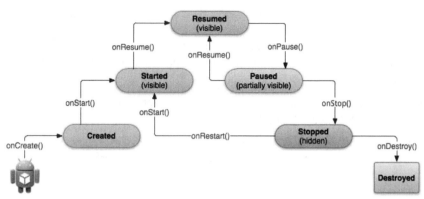

FIGURE 3-16: Android Activity lifecycle

Source: Figure 3.16 is reproduced from work created and shared by the Android Open Source Project and used according to terms described in the Creative Commons 2.5 Attribution License. (`https://developer.android.com/training/basics/activity-lifecycle/starting.html`)

To understand Android applications. you must understand the Activity lifecycle. Understanding the lifecycle is crucial to designing a better application while complying with the standards of Android applications.

> **NOTE** *Android activities are like the pages of a web site: They are assigned a specific task to present and users interact with them. The user switches between activities, going from one to another and back to the previous activity.*

The following list describes the Activity lifecycle states with Activity methods.

➤ onCreate—In the creation state, activities load the static user interface elements and allocate system resources such as the display device, network device, and other resources required for the application.

➤ onStart—After creation by the onCreate method, an activity enters the started state. It becomes visible and ready to handle user interaction.

➤ onResume—In the resumed state, all actions are performed, such as user input, drawing new user interface elements according to user input, and so on.

➤ onPause—In the paused state, the application is taken to the background, or another application is started and brought to the front.

➤ onStop—In the stopped state, the application is totally invisible and in a sleeping state.

➤ onDestroy—The destroyed state means that all processes are killed and all device resources released.

Activities handle drawing the user interface, user inputs, and response to user input. For example, when you launch an application, it launches the main activity to draw to the device's screen the first user interface defined in the layout XML file. It also handles touch inputs and responds to user interaction.

User-defined activities derive from the Android API's super class `Activity` in the `android.app` package. There are also other derived classes, which you can inherit to develop an Activity module for your application. `ListActivity` and `AppCompatActivity` are two Java classes derived from `Activity`. They provide extensions to configure your `Activity` class without requiring reimplementation of features you require.

The `Activity` super class inherits the activity lifecycle methods and application context fields that you can override to customize your own Activity for the application.

Listing 3-1 is from Android Studio's Empty Activity template. It overrides the `onCreate` method of the `Activity` super class to initialize the activity's `activity_main` user interface and displays "Hello World!"

LISTING 3-1: Empty Activity template code

```
public class MainActivity extends AppCompatActivity {

    @Override
    protected void onCreate(Bundle savedInstanceState) {
        super.onCreate(savedInstanceState);
        setContentView(R.layout.activity_main);
    }
}
```

The Intent Event Handler

Starting an application activity requires you to create an event handler called `Intent`. The `Intent` object is accessible through the `android.content` package. You can create a new `Intent` handler as follows:

```
Intent intent = new Intent();
```

`Intent` is a data structure that holds the description of the action to be performed, such as starting a new activity, application, service, and so on. The following code snippet creates a new `Intent` to start the `NextActivityClass` activity.

```
Intent intent = new Intent(this, NextActivityClass.class);
```

When a new application starts, the main activity becomes the first screen you see. The `Intent` abstract class is used by the Android system to launch the main activity. `Intent` is also used to start another activity within an activity.

Adding Template Activities to Android Projects

Android Studio provides several template activities for developers to seamlessly start development. In addition to providing a starting point for developing your application, they help ensure compliance with the Android application development guidelines.

In this section, you add two template activities: the Blank Activity and the Tabbed Activity. (You added the Empty Activity during project creation.)

Adding a Blank Activity

Let's add a Blank Activity template to your project. The Blank Activity template consists of two components in addition to the Empty Activity template.

Adding an activity template is an easy process in Android Studio. You can add a Blank Activity by performing the following steps.

1. Right-click on the project and select New ⇨ Activity ⇨ Blank Activity from the list of available activities, as shown in Figure 3-17. Alternatively, from the Android Studio menu, select New ⇨ Activity ⇨ Blank Activity. Both methods list the same activity templates from which to choose.

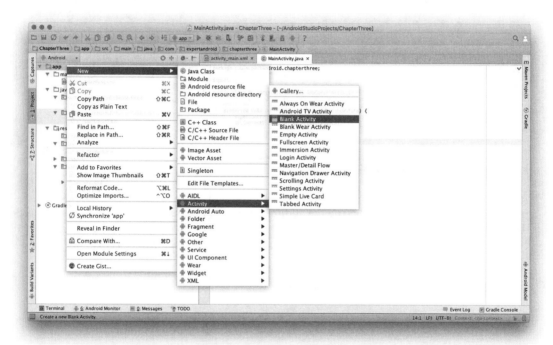

FIGURE 3-17: Adding a new activity template

2. Customize the new activity in the Customize the Activity window that displays (see Figure 3-18). Enter the name, title, UI layout name, and the Java package name you want to add to your new Activity class.

> **NOTE** *As you can see in Figure 3-18, you can choose that the activity be your launcher activity. You can also select the Use a fragment option to load content instead of the default, a floating button on the bottom-left corner. Finally, you can set the hierarchical parent activity, which will make this new activity directly navigate to its parent with an Up button.*

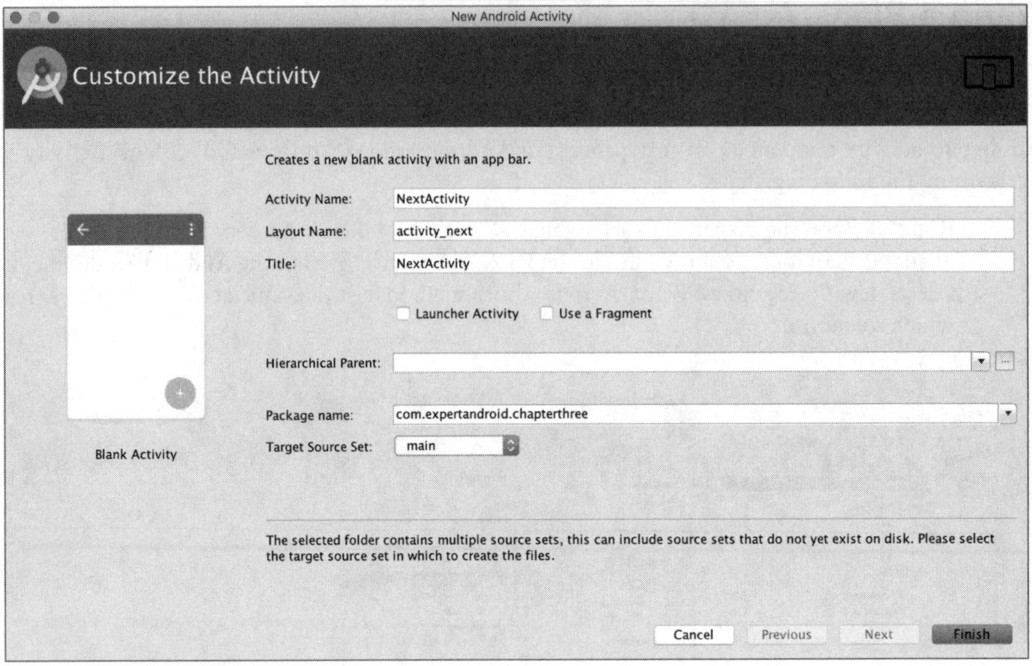

FIGURE 3-18: Activity template customization

After clicking Finish (without making the additional configurations mentioned in the previous note), the new activity's Java class, named `NextActivity.java`, is created in the `com.expertandroid` `.chapterthree` package. Listing 3-2 shows the auto-generated code from this example.

LISTING 3-2: Blank Activity template code

```java
public class NextActivity extends AppCompatActivity {

    @Override
    protected void onCreate(Bundle savedInstanceState) {
        super.onCreate(savedInstanceState);
        setContentView(R.layout.activity_next);
        Toolbar toolbar = (Toolbar) findViewById(R.id.toolbar);
        setSupportActionBar(toolbar);

        FloatingActionButton fab = (FloatingActionButton) findViewById(R.id.fab);
        fab.setOnClickListener(new View.OnClickListener() {
            @Override
            public void onClick(View view) {
                Snackbar.make(view, "Replace with your own action",
                        Snackbar.LENGTH_LONG)
                        .setAction("Action", null).show();
            }
        });
    }
}
```

As you can see in this source code, unlike the Empty Activity, the blank template includes a `toolbar` object and loads a layout to initialize its view on the phone or tablet.

In addition to the `toolbar` object, a `FloatingActionButton` object is defined for the Blank Activity and an action method to show a notification with Android API's `Snackbar` object has been defined by default with the template.

You cannot directly launch the Empty Activity you added earlier in the chapter. However, as the first activity you added, it is still the main launcher activity. As a result, you can either set an `intent` object inside the launcher activity to start the `NextActivity` class you just created, or you can set the `NextActivity` as the launcher activity when you add it. The second option makes it easier to display the Blank Activity onscreen immediately, as shown in Figure 3-19.

Together with the Java class, the template generates two XML files. Because we named the activity NextActivity, one class is named `activity_next.xml` and the other is `content _next.xml`.

The `activity_next.xml` file defines the main layout user interface elements, including a tool-bar widget (`AppBarLayout`) and a button widget (`FloatingActionButton`). `content_next.xml` is the layout defined to include user interface elements in the blank area.

FIGURE 3-19: Empty Activity template on an Android device

It has been defined as a `RelativeLayout` to configure and design the main user interface for the activity.

NOTE *Widgets are the user interface classes of Android API. All unique elements on the screen are derived from the widget super class. The following sections go into greater detail about UI elements.*

Listing 3-3 shows the `activity_next.xml` file for this example, without any modification.

LISTING 3-3: Blank Activity XML layout template

```xml
<?xml version="1.0" encoding="utf-8"?>
<android.support.design.widget.CoordinatorLayout
    xmlns:android="http://schemas.android.com/apk/res/android"
    xmlns:app="http://schemas.android.com/apk/res-auto"
    xmlns:tools="http://schemas.android.com/tools"
    android:layout_width="match_parent"
    android:layout_height="match_parent" android:fitsSystemWindows="true"
    tools:context="com.expertandroid.chapterthree.NextActivity">
```

```
    <android.support.design.widget.AppBarLayout
    android:layout_height="wrap_content"
        android:layout_width="match_parent"
        android:theme="@style/AppTheme.AppBarOverlay">
        <android.support.v7.widget.Toolbar android:id="@+id/toolbar"
            android:layout_width="match_parent"
            android:layout_height="?attr/actionBarSize"
            android:background="?attr/colorPrimary"
            app:popupTheme="@style/AppTheme.PopupOverlay" />
    </android.support.design.widget.AppBarLayout>
    <include layout="@layout/content_next" />
    <android.support.design.widget.FloatingActionButton android:id="@+id/fab"
        android:layout_width="wrap_content" android:layout_height="wrap_content"
        android:layout_gravity="bottom|end"
        android:layout_margin="@dimen/fab_margin"
        android:src="@android:drawable/ic_dialog_email" />
</android.support.design.widget.CoordinatorLayout>
```

All elements (the toolbar, the `content_next.xml` layout, and the `FloatingActionButton`) are
defined in the XML file inside another UI component named `CoordinatorLayout`. `content_next`
`.xml`. That file includes just a `RelativeLayout` with the required configurations but no other
elements. `RelativeLayout` is shown in Listing 3-4.

LISTING 3-4: Blank Activity Content Layout XML File

```
<?xml version="1.0" encoding="utf-8"?>
<RelativeLayout xmlns:android="http://schemas.android.com/apk/res/android"
    xmlns:tools="http://schemas.android.com/tools"
    xmlns:app="http://schemas.android.com/apk/res-auto"
    android:layout_width="match_parent"
    android:layout_height="match_parent"
    android:paddingLeft="@dimen/activity_horizontal_margin"
    android:paddingRight="@dimen/activity_horizontal_margin"
    android:paddingTop="@dimen/activity_vertical_margin"
    android:paddingBottom="@dimen/activity_vertical_margin"
    app:layout_behavior="@string/appbar_scrolling_view_behavior"
    tools:showIn="@layout/activity_next"
    tools:context="com.expertandroid.chapterthree.NextActivity">

</RelativeLayout>
```

Further development can be done inside the `NextActivity.java` file. You can override `onStart()`,
`onResume()`, and other Activity class functions to determine what this activity performs when it is
running. UI design and customization can be made in the `activity_next` and `content_next` XML
files, together with the Java class file.

In the next two sections, we briefly discuss fragments. Then, in the "Tabbed Activity" section, we
demonstrate an activity that includes a fragment.

Android Fragments

Fragments are like subactivities; they handle partial UI operations for a better user experience. Using fragments decreases the number of activities used inside the application and provides a smooth transition between user interface elements.

Using fragments simply divides the UI tasks defined for activities into subcomponents. Fragments allow developers to design a more compact application without the need to launch a large number of activities, which decreases the application stack.

Fragments are even more useful in large screen applications because you must make use of each part of the application without blocking the activity thread that handles the user interface.

Fragments are bound to a root activity to handle their jobs. For example, a typical mail application lists e-mail messages. When a user selects an e-mail to read, a fragment can be used to load only the e-mail content to the screen. If you use an activity instead of a fragment, you need to switch to the new activity and load the e-mail content to the activity; if you want to read another e-mail, you need to go back to the previous activity again, and so on. Fragments prevent this overload.

Fragments make an application respond faster and provide a more continuous user experience. Fragments can be reused within other activities, which helps reduce the number of fragment objects to be used.

Understanding the Fragment Lifecycle

Android fragments are like activities within an activity, so they have a similar lifecycle with additional bindings to the root activity they run in. Figure 3-20 shows the fragment lifecycle.

As described in detail in the previous section, fragments are like activities. A fragment has all the lifecycle states of an activity. In addition to the activity's lifecycle states, fragments have additional states to identify the activity they will be running in since a fragment's lifecycle is dependent on its parent activity. These additional lifecycle states are described in the following list.

➤ onAttach—The fragment's association with the root activity

➤ onCreateView—Creates and returns the view hierarchy associated with the fragment

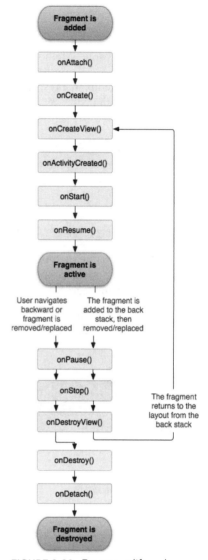

FIGURE 3-20: Fragment lifecycle

Source: Figure 3-20 is reproduced from work created and shared by the Android Open Source Project and used according to terms described in the Creative Commons 2.5 Attribution License. (https://developer.android.com/guide/components/fragments.html)

➤ onActivityCreated—Method called when the activity is ready after the onCreate state

➤ onDestroyView—Method called when the fragment frees resources and views

➤ onDetach—Method called when the fragment is no longer associated with the activity

The activity template presented in the following section is the Tabbed Activity, which has fragment definitions included.

> **NOTE** *The Tabbed activity template is not the only template that includes a fragment. The Master/Detail Flow template also includes fragment definitions.*

Adding a Tabbed Activity

The Tabbed Activity template is a complex activity template with fragments. The Tabbed Activity is also a very popular activity type that is used in many productive applications to switch between states of the application.

To add a Tabbed Activity, you can repeat the steps you followed to add a new Blank Activity earlier in this chapter, but the Tabbed Activity has a different configuration window, as shown in Figure 3-21.

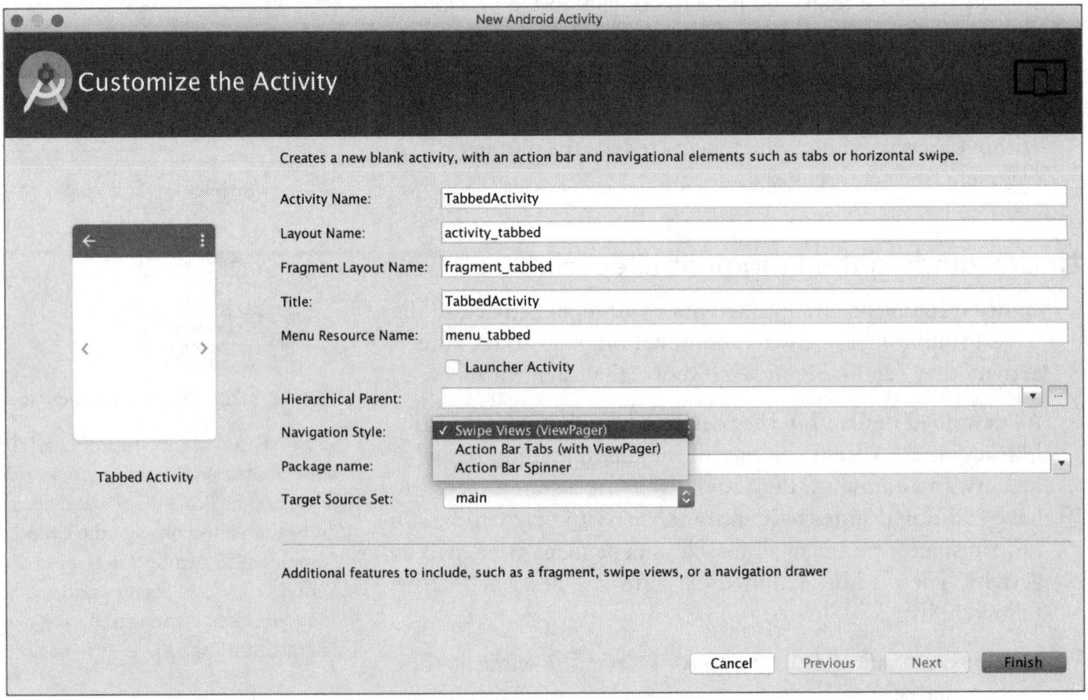

FIGURE 3-21: Tabbed Activity customization window

Figure 3-21 shows more fields to configure than the Blank Activity.

1. Named this activity as TabbedActivity, which auto configures the other fields, adding `_tabbed` to the Layout Name, Fragment Layout Name, and Menu Resource Name. The Title becomes TabbedActivity, as shown in Figure 3-21.

> **NOTE** *As with the Blank Activity, you have an option to select this activity as a LauncherActivity.*

2. Enter our project's default, `com.expertandroid.chapterthree`, as the package name.

The Navigation Style list offers three options; Swiping Views, Action Bar Tabs, and Spinner. These determine the style for switching between contents of the activity, such as changing between videos by clicking on a tab or using a swiping movement. By selecting the navigation style you can customize the UI.

When you change from one style option to another, you can see a preview at the left of the window shown in Figure 3-21. In this example, we do not recommend selecting a particular style because we want to show the results for each of them.

Changing the navigation style affects the UI widgets and views styles as well as the template's Java code. Swipe Views and Action Bar Tabs navigation styles generate similar code and XML files but the Action Bar Spinner is little different than the others because the method for navigation between fragments changes. The following list explains these three navigation styles:

➤ **Swipe Views**—Swipe screen to change between fragments.

➤ **Action Bar tabs**—Touch the tabs on the menu to switch between fragments.

➤ **Action Bar Spinner**—Use a drop-down menu on the toolbar to switch between fragments.

The Tabbed Activity is created from the super class `AppCompatActivity: public class TabbedActivity extends AppCompatActivity`.

The generated `TabbedActivity.java` template's code is longer and has additional methods to the Blank Activity because it has to account for swipes between fragment pages.

The Tabbed Activity template has two private fields for pager activity. The pagers, shown in the following code, handle switching between fragments.

```
private SectionsPagerAdapter mSectionsPagerAdapter;
private ViewPager mViewPager;
```

> **NOTE** *These pagers are not generated if the Action Bar Spinner navigation style has been selected. The Action Bar Spinner generates a drop-down list that helps the application switch between fragment objects.*

The onCreate function is similar to the same function in the Blank Activity template except there are initializations for paging operations. The onCreate function for the Tabbed Activity is shown in Listing 3-4.

LISTING 3-4: TABBED ACTIVITY TEMPLATE ONCREATE METHOD CODE

```java
@Override
protected void onCreate(Bundle savedInstanceState) {
    super.onCreate(savedInstanceState);
    setContentView(R.layout.activity_tabbed);

    Toolbar toolbar = (Toolbar) findViewById(R.id.toolbar);
    setSupportActionBar(toolbar);
    // Create the adapter that will return a fragment for each of the three
    // primary sections of the activity.
    mSectionsPagerAdapter = new SectionsPagerAdapter(getSupportFragmentManager());

    // Set up the ViewPager with the sections adapter.
    mViewPager = (ViewPager) findViewById(R.id.container);
    mViewPager.setAdapter(mSectionsPagerAdapter);


    FloatingActionButton fab = (FloatingActionButton) findViewById(R.id.fab);
    fab.setOnClickListener(new View.OnClickListener() {
        @Override
        public void onClick(View view) {
            Snackbar.make(view, "Replace with your own action",
                Snackbar.LENGTH_LONG).setAction("Action", null).show();
        }
    });
}
```

mSectionsPagerAdapter is set to FragmentManager, which interacts with the fragments associated with the activity. mViewPager is set to the container that includes the fragment's XML layout.

Dealing with fragments requires defining additional classes and functions for the activity. One class with its members is required to handle paging the activity and navigation between fragments: public classSectionsPagerAdapter extends FragmentPagerAdapter. (The constructor for this class is public SectionsPagerAdapter(FragmentManager fm) { super(fm); }.) The following functions help FragmentManager to get the page and to set the page title in the activity:

➤   public Fragment getItem(int position)

➤   public int getCount()

➤   public CharSequence getPageTitle(int position)

Then there is the Fragment class need to handle the views and states of the fragment. Listing 3-5 shows the Fragment class, which is auto-generated inside the TabbedActivity.java class and overrides the onCreateView() function to initialize the fragment's layout XML file.

LISTING 3-5: Tabbed Activity template fragment class code

```java
public static class PlaceholderFragment extends Fragment {
    /**
     * The fragment argument representing the section number for this
     * fragment.
     */
    private static final String ARG_SECTION_NUMBER = "section_number";

    /**
     * Returns a new instance of this fragment for the given section
     * number.
     */
    public static PlaceholderFragment newInstance(int sectionNumber) {
        PlaceholderFragment fragment = new PlaceholderFragment();
        Bundle args = new Bundle();
        args.putInt(ARG_SECTION_NUMBER, sectionNumber);
        fragment.setArguments(args);
        return fragment;
    }

    public PlaceholderFragment() {
    }

    @Override
    public View onCreateView(LayoutInflater inflater, ViewGroup container,
                             Bundle savedInstanceState) {
        View rootView = inflater.inflate(
            R.layout.fragment_tabbed, container, false);
        TextView textView = (TextView)
            rootView.findViewById(R.id.section_label);
        textView.setText(getString(R.string.section_format,
            getArguments().getInt(ARG_SECTION_NUMBER)));
        return rootView;
    }
}
```

When you launch the Tabbed Activity template, you see the UI on the device screen, as shown in Figure 3-22, if the Swipe Views navigation style is selected.

Figure 3-23 shows the screenshots taken for the other navigation styles: Action Bar Tabs and Action Bar Spinner.

Adding activity templates is easy with some practice, but the real work starts after you select the template that's suited to your application design. If your application will load content on the UI at each change, using a fragment is a good choice. If one screen on your application will be tasked to do a specific job, and you would not require a dynamic update on the UI content, you can add an activity without any fragments.

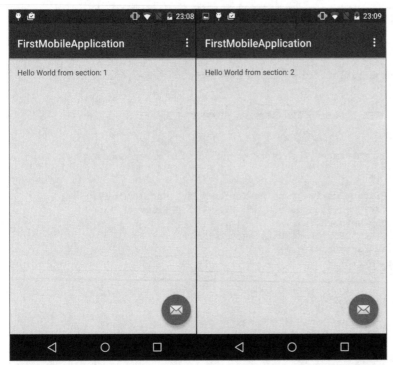

FIGURE 3-22: Tabbed Activity template screenshot with Swipe Views navigation style

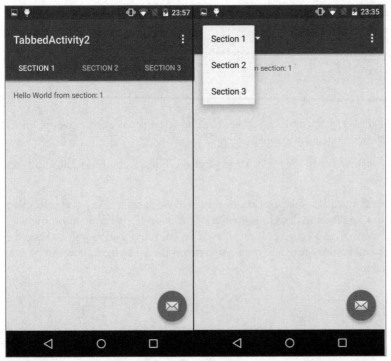

FIGURE 3-23: Action Bar Tabbed and Spinner UI screenshots for the Tabbed Activity template

Selecting templates for your application takes some effort and makes you focus on the design and development process. After your template and activity creation, you can work on the detailed implementation of features you want to add.

Activity templates save you from having to initialize your activity in the `AndroidManifes.xml` file because you create layout files in the right directory and perform initialization directly in the code. You save time with these great features of Android Studio.

In addition to Activity templates, Android Studio has templates for services, which are another main building block of Android applications. In the next section, you learn how to add service templates for your application.

ANDROID SERVICES

Services can be defined as activities without a user interface that run in the background to perform long-running tasks. Android applications can bind to services to perform file IO or retrieve sensor data.

A music player application is a good example to demonstrate services. When you open a music player application, start playback, and then switch to another application, the music player continues to play music in the background by using a service component that can stay active without a UI.

Android Studio provides two basic service templates for developers to add to their application. One is the standard Android Service class and the other one is the Intent Service.

Services usually handle tasks within the application's main thread; they are designed to be short running. However, `IntentService` handles long running tasks; they need to be designed as a separate thread than the main application thread.

Adding a Service Template with Android Studio

To add a service template to your application, follow the steps for adding an activity template, but click New ➪ Service ➪ Service or Service (`IntentService`).

Because services don't have a user interface, the new service wizard just asks for the service name and additional check buttons to configure service.

Figure 3-24 shows the window when you add a service.

There are two check buttons to configure Android Manifest initialization:

➤ **Exported**—When checked, other applications can invoke or interact with the service.

➤ **Enabled**—When checked, the system service can be instantiated by the system.

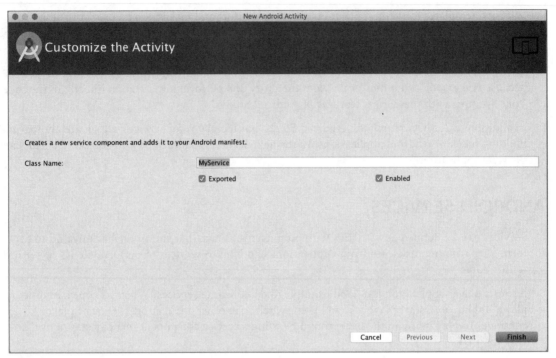

FIGURE 3-24: Customizing the service class name

The template source code overrides only the onBind method, which is used when another component wants to bind to the service (see Listing 3-6).

LISTING 3-6: Android Studio service template code

```
public class MyService extends Service {
    public MyService() {
    }

    @Override
    public IBinder onBind(Intent intent) {
        // TODO: Return the communication channel to the service.
        throw new UnsupportedOperationException("Not yet implemented");
    }
}
```

If you selected IntentService when you added the service template, you will see the window shown in Figure 3-25. Here you name the service and can check the option to add a static start method for the IntentService.

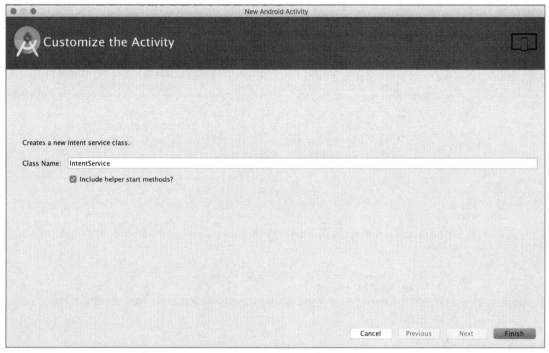

FIGURE 3-25: IntentService customization window

IntentService handles asynchronous tasks such as incoming requests from other components. For this reason, the IntentService class differs from the Service class. The template includes components related to asynchronous tasks you can customize as needed (see Listing 3-7).

LISTING 3-7: IntentService Template Code

```java
public class MyIntentService extends IntentService {
    // TODO: Rename actions, choose action names that describe tasks that this
    // IntentService can perform, e.g. ACTION_FETCH_NEW_ITEMS
    public static final String ACTION_FOO =
"com.expertandroid.firstmobileapplication.action.FOO";
    public static final String ACTION_BAZ =
"com.expertandroid.firstmobileapplication.action.BAZ";

    // TODO: Rename parameters
    public static final String EXTRA_PARAM1 =
"com.expertandroid.firstmobileapplication.extra.PARAM1";
    public static final String EXTRA_PARAM2 =
"com.expertandroid.firstmobileapplication.extra.PARAM2";

    public MyIntentService() {
        super("MyIntentService");
    }
```

```
    @Override
    protected void onHandleIntent(Intent intent) {
        if (intent != null) {
            final String action = intent.getAction();
            if (ACTION_FOO.equals(action)) {
                final String param1 = intent.getStringExtra(EXTRA_PARAM1);
                final String param2 = intent.getStringExtra(EXTRA_PARAM2);
                handleActionFoo(param1, param2);
            } else if (ACTION_BAZ.equals(action)) {
                final String param1 = intent.getStringExtra(EXTRA_PARAM1);
                final String param2 = intent.getStringExtra(EXTRA_PARAM2);
                handleActionBaz(param1, param2);
            }
        }
    }

    /**
     * Handle action Foo in the provided background thread with the provided
     * parameters.
     */
    private void handleActionFoo(String param1, String param2) {
        // TODO: Handle action Foo
        throw new UnsupportedOperationException("Not yet implemented");
    }

    /**
     * Handle action Baz in the provided background thread with the provided
     * parameters.
     */
    private void handleActionBaz(String param1, String param2) {
        // TODO: Handle action Baz
        throw new UnsupportedOperationException("Not yet implemented");
    }
}
```

ADD ASSETS FOR ANDROID PROJECT

While developing an application, you need to include resource files such as the logo art for your application, photos, figures, custom sounds, music, videos or animations. These files are the *assets* the application loads while running.

Each asset type should be added to its own directory to comply with the Android development standards. All assets are stored in the res folder. Initially, Android Studio creates only the drawable and mipmap-xdpi folders to store resources other than XML files.

> **NOTE** *This section covers how to add assets to your project; working with assets is covered in the following chapters.*

Switch to the traditional Project View to better see the res folder content, as shown in Figure 3-26.

Adding Images Assets

Images that you plan to use as a background on your project, or as a background for your buttons and so on, can be in many formats, such as PNG or JPEG. Image resources should be stored in the mipmap-xx folders (see Figure 3-26).

If your application will be running on multiple devices with different screen resolutions, you should add images to the corresponding resolution folder in your project directory and store the same image at different resolutions in the appropriate folder from the following list:

➤ **mipmap-hdpi**—High density images

➤ **mipmap-mdpi**—Medium density images

➤ **mipmap-xhdpi**—Extra high density images

➤ **mipmap-xxhdpi**—Extra extra high density images

➤ **mipmap-xxxhdpi**—Extra extra extra high density images

To add image resources to your project, follow these steps:

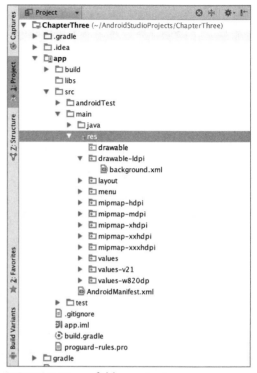

FIGURE 3-26: res folder content

1. Either drag-and-drop the file onto your project or right-click on the project and select Image Asset from the menu, as shown in Figure 3-27.

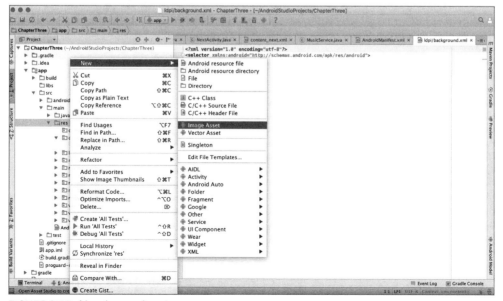

FIGURE 3-27: New Image Asset menu

2. Click Image Asset to open a new window where you can select the image size, type, and other properties to let Android Studio to customize the image, and load the image into all related folders easily as shown in Figure 3-28.

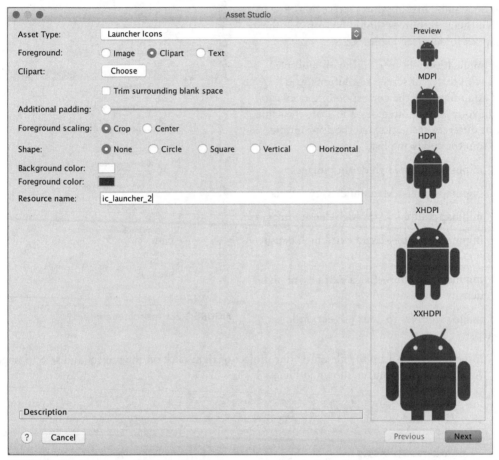

FIGURE 3-28: Image Asset Studio window

3. After completing the Asset Studio options, click Next. The Image Asset Wizard will show a summary of where the new resource will be placed (see Figure 3-29).

Adding Sound Assets

Sound assets are placed in subfolders in the raw directory. The raw directory is not auto-generated by Android Studio. If you would like to add a sound asset, create a subdirectory in the res folder by right-clicking on the res folder in Project View, and then select New ⇨ Folder. That opens a new window where you can name the new folder. Type **raw** and you are done adding the sound asset folder.

FIGURE 3-29: Image Asset Studio summary

> **NOTE** *These instructions follow the standard naming conventions but are not written in stone. You may create your own folder any time to store resources you need.*

Later you can add sounds by either dragging and dropping or copying and pasting to the Project View.

Adding Video Assets

Videos are also stored in the raw folder. Create subfolders for videos as you do for audio files (see the preceding section). Videos are not reliable because they are large files that can be expensive to load and play on a mobile device, so including videos is not preferred practice, although you may want to add a short animation of 2–5 seconds. That may create a better experience for your users.

ADDING XML FILES TO AN ANDROID PROJECT

You define static strings in XML files; these files contain static information to be encoded by an application at runtime. You sometimes need to define a list of strings in an XML file instead of declaring them in the Java source code. In fact, it is a very bad practice to define static strings in the Java source code, and Android advises developers to define static strings in the XML files. Figure 3-30 shows where strings are defined for the ChapterThree application.

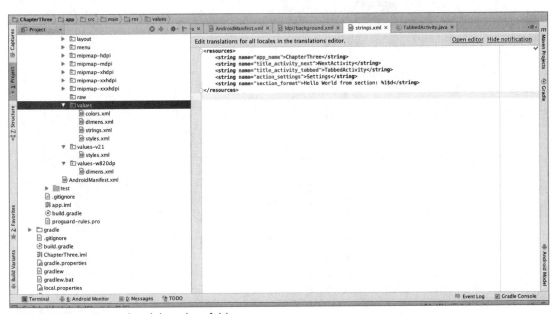

FIGURE 3-30: string.xml and the values folder content

XML files also define user interface modules, UI layouts, and styles to be reused in UI development. In order to add a new XML values file (not a layout file), right-click the values folder and select New ➪ XML ➪ Values XML File.

Since we will work on layouts in Chapter 5, we skip adding layout XMLs for this section. XML files also helps you to make an application more configurable, improve the UI, and implement more reusable code for your Android application.

ANDROID MANIFEST FILE

The Android manifest file is like an Android application's signature for the Android operating system. It defines the starting activity and other activities the application includes in the package, such as services, the application's name, the Android SDK version, and required permissions needed to access data and sensors on the device.

The Android manifest informs the Android operating system of possible processes and threads that can be generated for the application along with activity and service descriptions. The manifest informs the system that the application will get access to certain sensor devices and networks, will gather location information, and so on.

You might have noticed that some installed applications on your Android device ask for permission to access the Internet, your location, contacts, camera, and so on. These permissions are defined in the Android Manifest file.

The Android manifest file can be accessed from the Android manifest folder. Each time you define an activity or service manually, you should enter the required info into the `AndroidManifest.xml` file as well.

Listing 3-8 is the Android Manifest file of our application with the two activities and one service added in this chapter. You may notice that we also have requested Internet access with the `permission` tag. It can be found under the 'manifests' directory in the Android Project View.

LISTING 3-8: ANDROIDMANIFEST.XML SAMPLE

```xml
<?xml version="1.0" encoding="utf-8"?>
<manifest xmlns:android="http://schemas.android.com/apk/res/android"
    package="com.expertandroid.chapterthree" >

    <uses-permission android:name="android.permission.INTERNET"></uses-permission>

    <application
        android:allowBackup="true"
        android:icon="@mipmap/ic_launcher"
        android:label="@string/app_name"
        android:supportsRtl="true"
        android:theme="@style/AppTheme" >
        <activity android:name=".MainActivity" >
            <intent-filter>
                <action android:name="android.intent.action.MAIN" />
                <category android:name="android.intent.category.LAUNCHER" />
            </intent-filter>
        </activity>
        <activity
            android:name=".NextActivity"
            android:label="@string/title_activity_next"
            android:theme="@style/AppTheme.NoActionBar" >
        </activity>
        <activity
            android:name=".TabbedActivity"
            android:label="@string/title_activity_tabbed"
            android:theme="@style/AppTheme.NoActionBar" >
        </activity>

        <service
            android:name=".MusicService"
            android:enabled="false"
            android:exported="false" >
        </service>
```

```
      </application>
</manifest>
```

If you were using the Eclipse IDE previously, you were able to add tags with the help of UI components. You may struggle to find that capability in Android Studio; you should instead enter the new entry manually. Android Studio's IntelliSense feature will help you to auto-complete the entry in the Android Manifest file.

ANDROID MODULES

Modules are additional software components for the projects being developed. If you have created a project only for Phone and Tablet devices, initial application is the first module. Modules you can add include Android Library, Wear, TV, Glass, Phone and Tablet, and so on. Figure 3-31 shows all modules available to add on to your project.

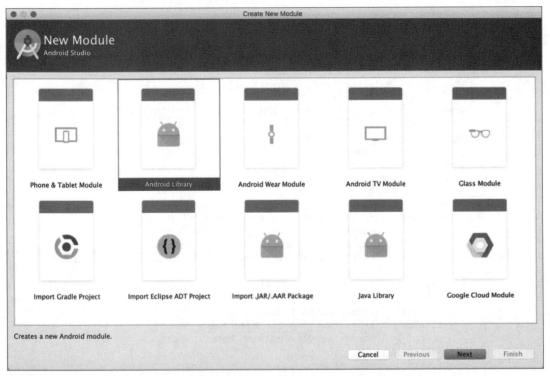

FIGURE 3-31: New Module selection window

Let's add an Android Library to our project. In the window shown in Figure 3-32, Android Studio helps you configure your module by first asking for the module name.

Create New Module

Phone & Tablet Module
Configure your new module

Configure your new module

Application/Library name:	MyAndroidLibrary
Module name:	myandroidlibrary
Package name:	com.expertandroid.myandroidlibrary Edit
Minimum SDK	API 23: Android 6.0 (Marshmallow)

Cancel   Previous   Next   **Finish**

FIGURE 3-32: Creating a new module

After you add the module, you will see a second directory in the Android Project View (see Figure 3-33).

You can add a new Java class to develop your own library to ease development of and reuse on other applications.

Modules have independent build configurations, and after the build process they also produce their own binaries.

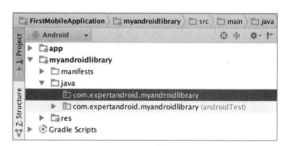

FIGURE 3-33: Project View after a new module is created

SUMMARY

In this chapter, we worked on creating the main building blocks of Android application projects using Android Studio. The main building blocks of Android applications are activities, services, assets, XML files, the Android Manifest file, and modules.

Brief information about building blocks has been provided to help you better understand what you are dealing with so you can follow the best practices of Android Studio application development.

Android Studio In Depth

The previous three chapters provided an introduction to environment setup and how to start a new project with Android Studio. This chapter dives into Android Studio's tools and best practices before digging into application development with Android Studio.

A good introduction to Android Studio tools will help you develop applications efficiently and enable you to take advantage of Android Studio as an IDE to ease the development process and optimize your application.

Throughout the chapter, we will review the Android Studio menu items and tools and answer questions you might have about why and when to use or access them. The chapter begins with menu items and then tackles shortcuts and tools used in Android Studio. Then you look at code refactoring with Android Studio and Live Templates. The chapter concludes with a focus on Android Studio's APK building utilities.

ANDROID STUDIO MENU ITEMS

If you are used to developing Android applications with Eclipse-based ADT, it may take a while to get to know Android Studio menu items.

Figure 4-1 shows all the menus: Android Studio (only in the Mac version), File, Edit, View, Navigate, Code, Analyze, Refactor, Build, Run, Tools, VCS, Window, and Help. The figure shows the Mac OS X version of Android Studio but the menus don't change much by OS.

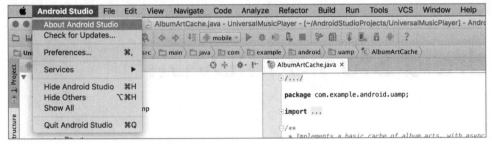

FIGURE 4-1: Menus on Mac OS X

> **NOTE** *Although menu items do not change between OSes, Mac OS and several Linux window managers tend to display menu items in the top bar whereas Windows keeps menu items in application windows.*

Familiarize yourself with the menu items within a desktop application so that you can easily utilize the application in a way that meets your specific needs and enables you to find a solution within the IDE itself for such tasks as adding a new file to your project, enabling additional application windows, configuring the application or checking for updates, and so on. The following sections examine Android Studio's menu items in more detail.

Android Studio

The Android Studio menu lists the main menu items, which provide access to updates and window management, along with the hide, quit, and show options, as shown in Figure 4-1.

> **NOTE** *The Android Studio menu isn't present in other OS versions, so most items listed in this menu are found in the File menu.*

The most important item in this menu is Preferences (Settings in Windows and Linux), which you may need to access many times to configure Android Studio.

The Preferences window gives you access to all the detailed Android Studio configurations: Appearance & Behavior, Keymap, Editor, Plugins, Version Control, Build, Execution, Deployment, Languages & Frameworks, and Tools, as shown in Figure 4-2.

> **NOTE** *The Keymap option is not covered in this section. We cover it in the section "Android Studio Shortcuts" later in the chapter.*

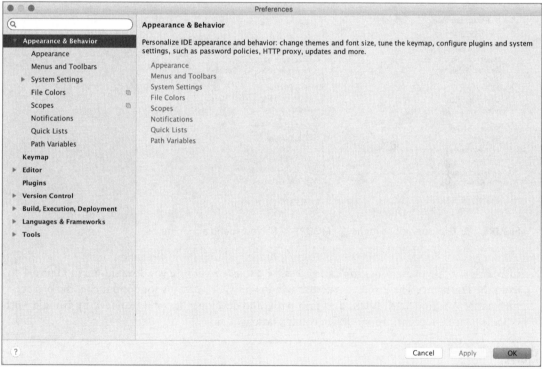

FIGURE 4-2: Preferences window

File

The File menu gives you access to file operations such as saving a file, adding a new file, selecting templates, opening existing files, and closing a project folder, as shown in Figure 4-3. It also includes file operations including setting the style of files in Android Studio, exporting any file as HTML, and setting Line Separators for the corresponding operating system to make a file readable by other operating systems.

In addition to the File menu giving access to file settings, it also has a New submenu (see Figure 4-4) with actions to create a new project and to import a project from local sources or from a version control system. Adding and importing a new module is also done from the New menu. Finally, can add a new file, directory, C++ Class, C/C++ Source or Header file with the actions defined in the New menu.

FIGURE 4-3: File operations menu   **FIGURE 4-4:** New menu action items

In addition to file operations, you can access the Project Structure window shown in Figure 4-5 from the File menu. The Project Structure window includes options for configuring the project, such as the Android SDK, NDK, and Java path, and developer services provided by Google, such as Google Sign-In authentication service configurations.

Edit

The Edit menu includes text operations such as copy, paste, and cut functionality, and so on, as shown in Figure 4-6. The following list provides a quick reference to the operations for copying and pasting during coding.

➤   **Copy Path**—Copies the full path of a recently opened file

➤ **Copy as Plain Text**—Copies any text without formatting.

➤ **Copy Reference**—Copies the reference to the file in the system. For example, if you select Copy Reference for a Java file, it will copy only the name of the Java class.

➤ **Paste from History**—Lists previously copied texts and so you can select from the history to paste them.

➤ **Paste Simple**—Pastes the last copied text without formatting.

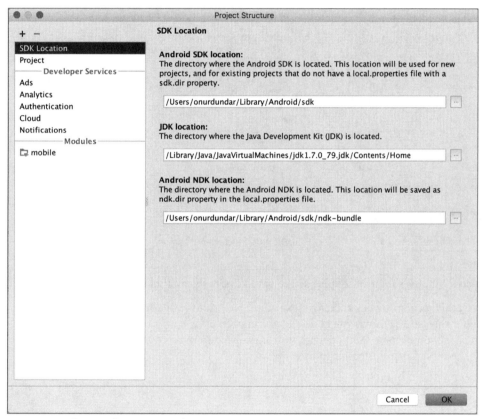

FIGURE 4-5: Project Structure window

The Edit menu options are not limited to basic copy and paste; there are more complex functions for searching and selecting under the Find submenu such as Find in Path and Replace in Path, Find, Duplicate Line, and Join Lines (see Figure 4-7).

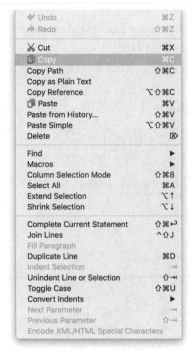

FIGURE 4-6: Edit menu items

FIGURE 4-7: Find menu items

There is also a Macros submenu that is used to save keystrokes to a macro for reuse.

In summary, this menu provides access to text editor functions available in most text editors. These items simplify the process of writing code and editing files. We recommend that you become familiar with shortcuts listed in the menu to make these functions easier to use.

View

The View menu, shown in Figure 4-8, provides easy access to functions for working with windows, configuring the editor spacing, adding line numbers, accessing the compare tool, and other similar functionalities of Android Studio.

Through the View menu, you can change the current view of Android Studio as well as enable or disable the Tool buttons, Toolbar, Status Bar, and Navigation Bar. In summary, this menu will help you change or customize the view of Android Studio's windows.

Navigate

The Navigate menu, shown in Figure 4-9, lists the actions you can take to navigate between files in the current open project. Practice

FIGURE 4-8: View menu items

with the navigation functions will make it easy to handle large projects with many files, classes, and long files.

You can add bookmarks to file lines so you can directly jump there during development, or navigate to a superclass by pressing Command+U on Mac or Ctrl+U in Windows. As shown in Figure 4-10, you can also see all the implementations of a superclass to enable you to navigate to a subclass.

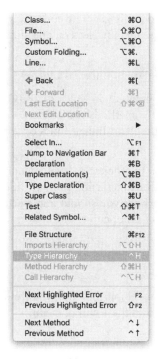

FIGURE 4-9: Navigate menu items

```
public abstract class BroadcastReceiver {
    private Pendin
    private boolea
                        Choose Implementation of BroadcastReceiver (13 found)
                        Anonymous in mAudioNoisyReceiver (com.example.android.uamp)
    /**                 Anonymous in mConnectivityChangeReceiver (com.example.android.uamp.ui)
     * State for a      Anonymous in onCreate() in MusicService (com.example.android.uamp)
     * by {@link B      AppWidgetProvider (android.appwidget)
     * while in {@      DeviceAdminReceiver (android.app.admin)
     * This allows
     * terminate;       MediaNotificationManager (com.example.android.uamp)
     * broadcast.       RestrictionsReceiver (android.service.restrictions)
     * thread of y      StatusReceiver in RemotePlaybackClient (android.support.v7.media)
     *                  VideoIntentReceiver (com.google.android.libraries.cast.companionlibrary.remotecontrol)
     * <p>Note on       VolumeChangeReceiver in LegacyImpl in SystemMediaRouteProvider (android.support.v7.medi
     * thread-safe      WakefulBroadcastReceiver (android.support.v4.content)
     * sure that y
     * the entire      zzb in CastRemoteDisplayLocalService (com.google.android.gms.cast)
     * for setting      zzb in zzg (com.google.android.gms.common.api)
     */
    public static
```

FIGURE 4-10: Implementation of the superclass BroadcastReceiver

All functionality provided in the Navigation menu will help you manage and find related files, folders, classes, declarations, and inherited classes easily and to understand the architecture of the software and avoid wasting time looking for a superclass's definition. Developing familiarity with these functions will definitely increase how much you like the IDE.

Code

The Code menu, shown in Figure 4-11, provides access to functions to easily generate code snippets, arrange code lines, add block comments, access Live Templates, add getters and setters, and similar activities.

Most of the functions are usable if you are in a Java file, but not if you are not in a Java file.

Let's look at two of this menu's options. Click Override Methods (Control+O on Mac or Ctrl+O in Windows) while in a Java class to open a new window that shows the methods you can override from inherited classes. If you want to re-indent your code, click the Auto-Indent Lines option (Control+Option+I on Mac or Ctrl+Alt+I in Windows).

Familiarizing yourself with the shortcuts on this menu will enable you to efficiently develop your code. For example, if you just inherited an interface, navigate to the Code menu and press the indicated shortcut for the Implement Methods option to see the list of methods you should implement for the current class. This saves the time that would otherwise be spent on documentation or navigating between superclasses.

Analyze

Analyzing your code with Android Studio is fun. Analysis gives you many insights into what to do next to improve the quality and stability of your code. Figure 4-12 shows the Analyze menu.

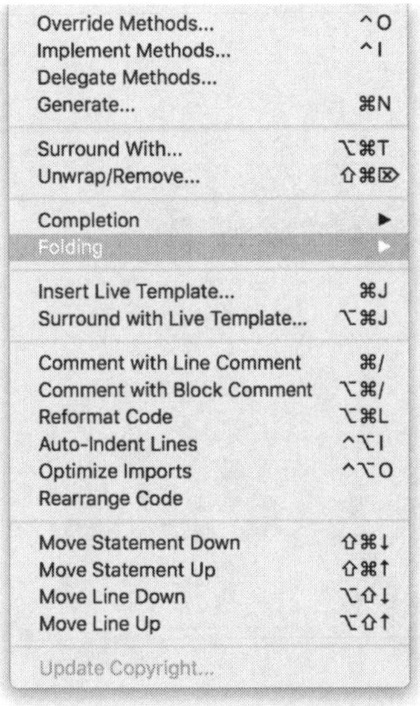

FIGURE 4-11: Code menu items

FIGURE 4-12: Analyze menu items

From the Analyze menu, let's inspect the UniversalMusicPlayer's `MediaNotificationManager.java` file. With the file open, select Inspect Code. This will open a new window at the bottom of the main editor and project pane to show the code maturity, possible bugs, and spelling and declaration suggestions, as you can see in Figure 4-13.

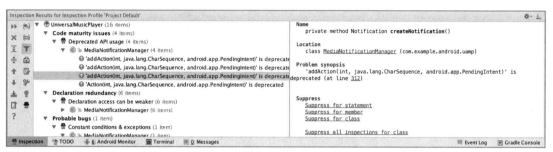

FIGURE 4-13: Inspection window

Refactor

Refactor menu tools provide developers easy access to certain refactoring operations, as shown in Figure 4-14.

To start refactoring, select a piece of the code in the file and then press Control+T on Mac or Ctrl+Alt+Shift+T in Windows to display the refactoring options, as shown in Figure 4-15.

Refactoring a large project might be difficult, but Android Studio's refactoring tools, when appropriate for your purpose, leads you to the right standards-compliant approach. The section "Code Refactoring with Android Studio" later in this chapter provides additional information on refactoring.

Build

The Build menu lists the commands to build the application and related configuration functions. These commands include APK generation of Android applications, signing APK files, editing dependencies, configuring build types, and selecting SDK versions, as you can see in Figure 4-16.

The Clean Project and Rebuild Project options are straightforward. They start clean and build functions as in other IDEs. Other options in this menu are mostly related to the output APK file configuration. We will cover these in greater detail in the APK generation and signing APK sections, later in the chapter.

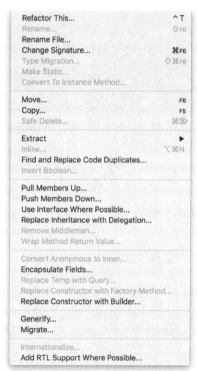

FIGURE 4-14: Refactor menu items

Run

The Run menu provides options for managing application process-ing such as running in debug mode or not, or accessing APK release configuration while running/debugging the application.

You can see all the shortcuts related to running and debugging your application from the menu. It will give you fast access during debug-ging. If your application is not in debugging mode, most of the menu items are disabled.

Tools

The Tools menu provides some tools not directly related to Android application development. You can create tasks to follow up with your development, generate JavaDoc, and create new scratch files to test some code snippets directly inside the project instead of creating a new Java or Android project, This menu is shown in Figure 4-17.

Finally, from the Tools menu you can launch the Android SDK Manager, AVD Manager, and Android Device Monitor.

FIGURE 4-15: Refactor options

Version Control System

The VCS (Version Control System) menu, shown in Figure 4-18, gives access to tools related to source control management. Android Studio helps developers to locally control the source code history, integrate with a selected version control system, and check out the source code from the remote repository.

Chapter 9 explores the use of Android Studio's version control system in greater detail.

FIGURE 4-16: Build menu items

Window and Help Menu

The Window menu helps you manage tabs and windows, includ-ing the management of each tool window's visibility type, such as docked, floating, and so on. You can also navigate between Android Studio windows if multiple Android Studio windows are open.

The Help menu gives you access to the Android Studio documenta-tion, where you can easily search for an action. If you want to look for the refactor or block comment shortcut, go to the Help menu and click Find Action (Shift+Command+A on Mac or Ctrl+Shift+A in Windows) to search for a specific action in Android Studio.

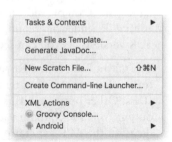

FIGURE 4-17: Android Studio Tools menu

ANDROID STUDIO SHORTCUTS

IDE shortcuts are important for any developer because an efficient use of an IDE starts with knowing how to easily refactor code, find any file or text in the project, or navigate to a required file by pressing a combination of keys on the keyboard. It is like living in a country and knowing all the public transportation alternatives or driving routes without spending any time searching the web. It can save you a lot of time.

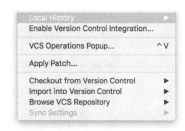

FIGURE 4-18: VCS menu items

Actually, you have seen many shortcuts for some operations in the figures in the previous sections. However, the menu items do not provide a comprehensive list of shortcuts—there are more than a hundred shortcut key mappings in Android Studio. You can see them all in the Keymap page of the Preferences window, as shown in Figure 4-19.

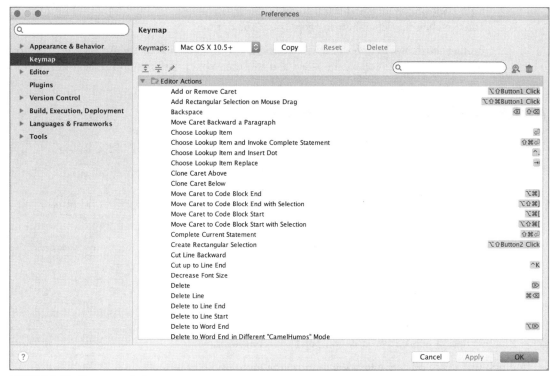

FIGURE 4-19: Keymap window

The Keymap window is useful not only for viewing the list of shortcuts, but also for editing, deleting, or adding shortcuts. It is always possible that existing mappings are not sufficient or not useful for you, so you can just edit them how you feel comfortable.

The following actions are available in Keymap:

➤ **Adding Shortcut**—Go to an empty action and right-click on it or click the pencil button above the list of Editor Actions in the Preferences window; then select between Add Keyboard Shortcut, Add Mouse Shortcut, or Add Abbreviation for the action.

To try this, select an action, and either from the right-click menu or after clicking the pencil button, select Add Keyboard Shortcut to open the window shown in Figure 4-20. With the cursor in the First Stroke area, press the key combination you want to assign to the action you selected. If the key combination is already assigned, you will see the conflict in the Conflicts area.

➤ **Deleting a shortcut**—When you want to delete the shortcut for an action, find the action in the Keymap list, right-click on the item and select Remove. Alternatively, you can click the pencil button again to see the options.

FIGURE 4-20: Enter Keyboard Shortcut window

➤ **Editing shortcut**—In order to edit a shortcut, you should click on the Add Keyboard Shortcut option as discussed earlier in this list and edit the stroke, as shown in Figure 4-17.

ANDROID STUDIO TOOL VIEWS

Android Studio provides a number of useful tools that help you take control and monitor the application development process efficiently. These tools are available from the Tool Windows option on the View menu. Figure 4-21 shows the list of tools: Messages, Project, Favorites, Run, Debug, Android Monitor, Structure, Version Control, Android Model, Build Variants, Capture Analysis, Capture Tool, Captures, Designer, Event Log, Gradle, Gradle Console, Maven Projects, Palette, Terminal, and TODO.

Notice that in Figure 4-21 some of the tools are disabled. These tools will be available when the development context is available for the use of the tool. For example, Designer and Palette get active when you start designing user interfaces. When you start debugging your application, the Debug option will be available to open the Debug window.

A quick way to see all active views is to move the mouse pointer to the bottom-left corner of the main Android Studio window and hover over the square icon. That opens the Tools list shown in Figure 4-22 so you can select and open the tool you need.

FIGURE 4-21: Tools Window items

Available tools are displayed at the edges of the Android Studio window. You can activate them either from the list or by clicking on the buttons on the window's edges. Figure 4-23 shows that the Project and Terminal tools are activated. When active, their background color is darker.

> **TIP** *Clicking the bottom-left corner of the window will hide all the edges and thus hide the clickable tool's buttons from the edge.*

It is essential for a developer to know both programming and the IDE—in detail—to solve development issues immediately and create a stable application. We see these tools as an essential part of the Android development process.

The following subsections cover the details of the tools mentioned to this point. To help clarify the discussions of the tools that follow, let's open the UniversalMusicPlayer sample project to better show the options. (Opening a sample application was covered in Chapter 2.)

FIGURE 4-22: Accessing the tools shortcuts

Messages

The Messages tool, shown in Figure 4-24, is available by default; it will get active when there is a need to show messages to the developer such as errors, warnings, information, notes, and generic information about the build process, as shown in Figure 4-24.

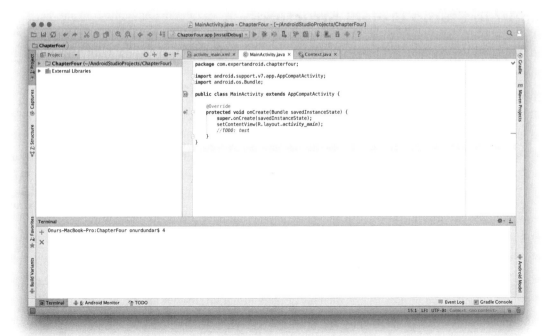

FIGURE 4-23: Activated tools

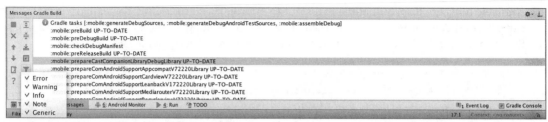

FIGURE 4-24: Messages tool window

In Figure 4-24, the Messages window shows the output messages from a Gradle build. If you opened the sample application as suggested in the previous section, it may automatically start the build process and show the output from the Messages window.

The Messages tool helps you follow what other tools are doing. In showing the errors, information, and warnings for the process in the window, the tool reveals what is going on in the background. Otherwise, it would be hard to follow up with ongoing processes.

The Messages tool has additional features to export messages in a text file to share with your team, get help, and resolve issues quickly.

The buttons on the left-hand side of the window let you easily navigate between messages. That can be difficult because there will often be more than the ten lines of messages shown in Figure 4-24; you may encounter thousands of lines. The buttons help you to expand or filter the list, and even go to the source of the error when necessary.

Android Studio Project Structure

Android Studio's Project tool is responsible for listing your project's files. It helps you directly browse the files and folders of your project in a tree structure from within Android Studio.

Android view shows files and folders grouped under Android modules and scripts. Each Android Module's Java source, test, manifest, and resource files will be grouped under the module name. Scripts will be under a different group named Gradle Scripts as shown in Figure 4-25.

The file and folder view options shown in Figure 4-25 are useful for creating a custom view for developers working on separate parts of the project because they enable the developer to focus on the context of his or her area of development. Compared to Eclipse IDE, this feature helps you filter to a simplified view of the project instead of forcing you to see all the files; it eliminates most auto-generated files, which you generally don't need to view. You can interpret this as a simple separation of concerns by Android Studio.

The view options change the file and folder listings as follows:

> **Project**—Lists all files and folders located in the Android application's project folder, which select while you are populating the project. This view option also lists the external libraries used in the project.

➤ **Packages**—Lists the files with package classification. XML files, menu files, layout files, drawable files, and Java source files are classified under a list of folders to ease direct access to the package you are working on.

➤ **Scratches**—Shows the list of scratch files created and used.

➤ **Android** —Shows only an application's related files. Applications are shown as modules— mobile, wear, library, and so on. This view enables fast access to your application.

➤ **Project Files**—Presents the build, license, and source control management–related project files. These include the `.gitignore` and `local.properties` files, and the `.google` and `.ide` folders.

➤ **Problems**—Shows only files with problems. To test this feature, delete a semicolon from a line in a Java file. The file will then be shown in this view to direct you to the problem.

➤ **Production**—Shows the folders related to the production of the application, such as the build output folder and the configuration folder, as well as the final source files which affect the production version of the application.

➤ **Tests**—Shows test-related results.

➤ **Android Instrumentation Tests**—Presents the Android Instrumentation test files.

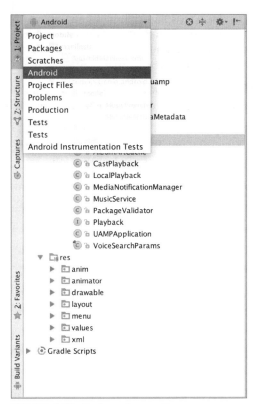

FIGURE 4-25: Project view options

Favorites

The Favorites tool, shown in Figure 4-26, provides fast access to your favorite folders and files, to bookmarked locations in files, and to toggled break points in the project.

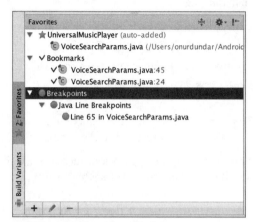

FIGURE 4-26: Favorites window

Adding a favorite is easy, just right-click a file in Project view or right-click an opened file's tab in the editor; then click Add to Favorites. The first time you do this, you will see only one group to add your file to in the favorites list. If you are following our example, it is UniversalMusicPlayer. You can also generate your own favorite group to classify your favorites under different lists.

Adding a new bookmark is also easy: Press F3 or F11 in Windows or use the Navigate menu's Bookmark option. A bookmark will be added to a line you select and the bookmark will be listed in the Bookmarks list in the Favorites window.

You can navigate between bookmarks by selecting the Show Bookmarks option from the Navigate ⇨ Bookmarks menu, or just press Command+F3 on Mac or Shift+F11 in Windows.

Breakpoints are added as favorites in the same way as Bookmarks. After you add a breakpoint in a file, it is listed in the Favorites window. In Figure 4-26, you can see that we added two bookmarks and one breakpoint to the sample project randomly.

Android Monitor

While developing your application, you will use the Android Monitor extensively (see Figure 4-27). With this tool, you can start and stop tools that monitor the use of resources on the Android device. The Android Monitor monitors the application during debugging or when it's running on the device. Processes monitored include GPU, memory, CPU, and network utilization.

The Android Monitor includes Android's logger tool, the Logcat utility. Figure 4-27 shows a typical log output.

Chapter 8 covers Android Monitor tools in detail.

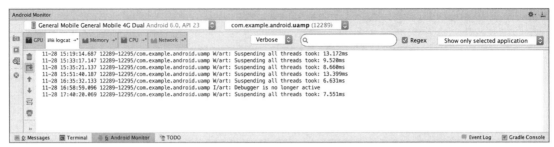

FIGURE 4-27: Android Monitor window

Structure

The Structure window lists the components of structured files such as Java, XML, and Gradle. Using this view, you can see all the methods, fields, and tags of XML files, and inherited members of a Java class. Instead of scrolling up and down, you can directly access a member of the file and work on it.

In our sample application, let's open a more complex class named MediaNotificationManager, placed under java sources, inside the com.example.android.uamp package. MediaNotification Manager inherits members from BroadcastReceiver to be able to list the details of the Java class, as shown in Figure 4-28.

In this window you can see the members that are inherited from BroadcastReceiver or Java's object class itself. Constant (final) or static members are identified by a lock symbol. In addition, (**m**) indicates the methods, (**f**) indicates the fields of the class, (**C**) shows the classes, (**I**) shows the interfaces, (**p**) indicates properties of class. This helps you understand any Java classes' design and architecture and assists with analyzing existing code. For example if you are navigating in a large project to understand the design of the classes, you can see the structures easily to understand how they were created.

You can also open XML files to see the details and relationship of the tags to each other.

Android Model

The Android Model view provides an easy access to an Android application member's initial values. These values do not contain the Java class members; the model includes the project build members, folder locations, compilation configurations, build type initializations, APK signing options, and so on.

The list is very long so it is hard to remember all the required fields to configure. Having this tool, which lets you see and review all the configuration's constant values, is very valuable. As shown in Figure 4-29, you can see all initializations.

This example shows only a single mobile application; if you have multiple modules in your project, the tool will show related configuration parameter values.

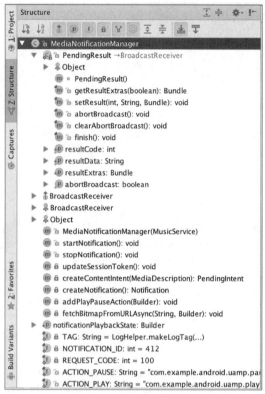

FIGURE 4-28: Structure view of a Java class

Gradle and Gradle Console

The Gradle and Gradle Console window tools show Gradle tasks in your project. When you open the Gradle window, you can execute any task independently and remove or add a new Gradle task to your project.

Gradle Console connects to Gradle builds, and this window shows whether the Gradle build tasks are successful or not. It is a simple console that shows only the output text.

We don't do any more changes to the default Gradle configuration and tasks in this chapter, because we will cover Gradle and Gradle Console in more detail in Chapter 6, which discusses the Gradle build system.

Run

The Run window is enabled when you run the application (as shown in Chapter 3) on a remote Android device or emulator. It lists the process and commands launched during the run phase such as APK installation on the device and the process ID of the application.

As shown in Figure 4-30, we launched the UniversalMusicPlayer application from our development machine to a remote Android phone.

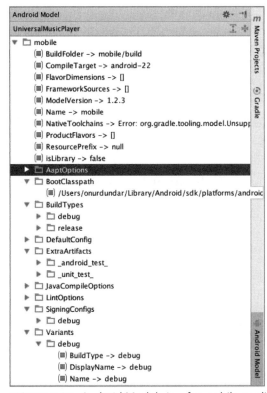

FIGURE 4-29: Android Model view for mobile applications

```
Run   mobile                                                                                    ✿ ⋅ ⊥
G  ↑    Target device: general_mobile-general_mobile_4g_dual-f2f6c6c5
        Installing APK: /Users/onurdundar/AndroidStudioProjects/UniversalMusicPlayer/mobile/build/outputs/apk/mobile-debug.apk
■  ↓    Uploading file to: /data/local/tmp/com.example.android.uamp
        Installing com.example.android.uamp
        DEVICE SHELL COMMAND: pm install -r "/data/local/tmp/com.example.android.uamp"
            pkg: /data/local/tmp/com.example.android.uamp
✕       Success
?
        Launching application: com.example.android.uamp/com.example.android.uamp.ui.MusicPlayerActivity.
        DEVICE SHELL COMMAND: am start -n "com.example.android.uamp/com.example.android.uamp.ui.MusicPlayerActivity" -a android.intent.action.MAIN -c android.intent.category
        Starting: Intent { act=android.intent.action.MAIN cat=[android.intent.category.LAUNCHER] cmp=com.example.android.uamp/.ui.MusicPlayerActivity }

        Connected to process 11607 on device general_mobile-general_mobile_4g_dual-f2f6c6c5
        |

 0: Messages    Terminal    6: Android Monitor    4: Run    TODO                        3 Event Log    Gradle Console
```

FIGURE 4-30: Run window

This tool also gives you the ability to stop, run, or re-run the application with the buttons shown at the left.

Debug

The Debug window is inactive by default and activated when you start debugging your application. After you click the debug button at the top of the Android Studio window or press Control+D on Mac or Shift+F9 in Windows to launch the application in debug mode, the Debug window appears at the bottom, as shown in Figure 4-31.

FIGURE 4-31: Debug window

In short, this window enables you to run an application step by step and see the variables. It also provides all required debugging ability to analyze and find the vulnerabilities of the application. (Detailed debugging properties are investigated in Chapter 8.)

Event Logs

The Event Logs tool prints events such as the Gradle build's start date and when you started an application's run session. It also prints errors or warnings that have occurred. The best feature of the Event Log tool is that it notifies users about these events and their start and finish times. Event logs help you understand the history of your development.

Terminal

The Terminal allows you to interact with the operating system's shell. It helps you do the required configuration and file manipulation, such as deleting, moving, and renaming files and folders in the current project directory without changing your context.

There are tools provided with Android SDK that do not have GUI support, so they are accessed via the command line.

As shown in Figure 4-32, when you open the Terminal window, you enter the current user's project's root folder.

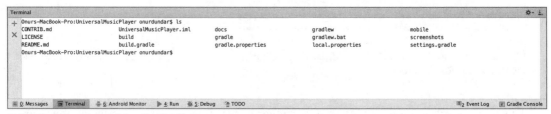

FIGURE 4-32: Android Studio Terminal

ANDROID STUDIO EDITOR

Android Studio comes with powerful and context-aware editors. Depending on what file format you are working with, Android Studio highlights, formats, indents, offers auto completion, uses color and text formatting, provides smart navigation, and, of course, includes all mandatory editing tools.

Code Assist

The killer feature of Android Studio editors is code completion assist. This feature has been a standard in similar IDEs such as Eclipse, Visual Studio, and NetBeans; however, Android Studio brings the assist to a completely new level.

For basic code completion, you can use Control+space on Mac or Ctrl+space in Windows. Although this works pretty much like the code completion offered by other IDEs, it is case-sensitive. For example, typing `Na` after hitting Control+space will list all methods that have that substring, such as `getName` and `setName`; typing `get` will list each method that starts with the string `get` and methods where the string `get` occurs somewhere in the method name.

Another great feature offered by Android Studio is Smart Type Completion. If you press Control+Shift+space on Mac or Ctrl+Shift+space in Windows for completion assist, Android Studio will filter the suggested item list with compatible return types. This offers smart and context aware code completion. If you hit Control+Shift+space after typing **Na,** which would be assigned to a `String` variable, you get only `getName` but not `setName`.

Commenting Out Code Blocks

Commenting out a piece of code might be the most used feature of an IDE. The key combinations Control+Shift+/ on Mac or Ctrl+Shift+/ in Windows will comment out the selected portion of the code with the proper syntax.

Moving Code Blocks

Moving code blocks without cut and paste can be very effective and helpful. Android Studio supports the following commands to help you to move the code around in the editor without cut and paste:

➤ **Move line up: Command+Shift+ up arrow on Mac (Ctrl+Shift+up arrow in Windows)**— Because the editor is context aware, if the line is broken into several lines, all related lines are moved.

➤ **Move line down: Command+Shift+ down arrow on Mac (Ctrl+Shift+ down arrow in Windows)**—Because the editor is context aware, if the line is broken into several lines, all related lines will be replaced.

➤ **Move line to the top: Command+Shift+ up arrow on Mac (Ctrl+Shift+ up arrow in Windows)**—Because the editor is context aware, it will not move a variable or method above the class declaration.

➤ **Move line to the bottom: Command+Shift+ down arrow on Mac (Ctrl+Shift+ down arrow in Windows)**—Because the editor is context aware, it will not move a variable or method outside the class parenthesis.

All options also work with selected blocks of code, enabling you to move large pieces of code blocks.

Navigating Inside the Editor

The capability to jump to the right place in the file you are editing can greatly reduce the time you spend navigating. Android Studio is very helpful and offers the following flexible options:

➤ Move cursor to the last editing position: Command+Shift+Delete on Mac (Ctrl+Shift+Backspace in Windows)

➤ Move cursor to the start of the current code block: Command+Option+[on Mac (Ctrl+Shift+[in Windows)

➤ Move cursor to the end of the current code block: Command+Option+] on Mac (Ctrl+Shift+] in Windows)

➤ Move cursor to the previous word: Option+left arrow on Mac (Ctrl+left arrow in Windows)

➤ Move cursor to the next word: Option+right arrow on Mac (Ctrl+right arrow in Windows)

➤ Select and move cursor to the previous word: Option+Shift+left arrow on Mac (Ctrl+Shift+left arrow in Windows)

➤ Select and move cursor to the next word: Option+Shift+right arrow on Mac (Ctrl+Shift+right arrow in Windows)

Refactoring

Refactoring is another area where Android Studio shines. From changing a variable or method name to extracting a block of code to a method, Android Studio offers many powerful refactoring options; they are covered later in this chapter, in the section "Code Refactoring in Android Studio."

Refactoring in Android Studio also checks other types of resources and performs string name checking to make sure the refactoring does not cause any compilations or runtime problems.

ANDROID STUDIO LIVE TEMPLATES

Live Templates are predefined code snippets that you can easily add to your code so you don't have to write the same code over and over. Live Templates are very useful when it comes to repeating a specific type of code block such as loops. For example, you may need to use a simple `for` loop many times in your code so adding a template would ease that process.

Live Templates may also be very useful if you want to create coding standards for your company. When you have a new developer in your company or team, you can just encourage him or her to use the predefined templates you already created while they work on new projects.

There are many predefined Live Templates already in Android Studio. In order to see, add, or remove Live Templates, navigate to Preferences, expand to Editor, and then click Live Template.

You will see an expandable list of names for the Live Template group, which includes the Live Templates shown in Figure 4-33.

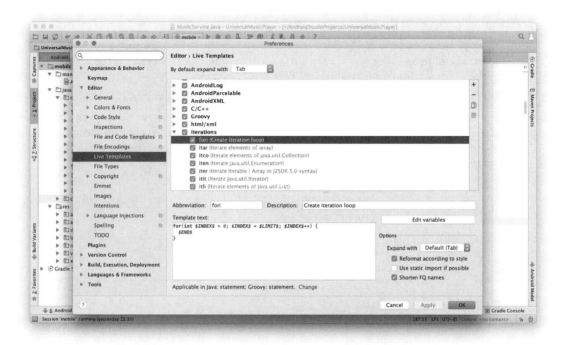

FIGURE 4.33: Live Template window

Live Templates are available for a wide variety of uses, such as comments for code, generic tags for Android XML, and so on. You can also customize them as needed during development.

Lists of Live Templates are available by expanding all the groups. As you may notice, Live Templates are very common code snippets with a high number of reuses at any point of your Java code. For example, `fori` is a predefined Live Template that creates a code snippet to create a basic `for` loop.

Inserting a Live Template

When you want to add any of the templates to your code, just type the name of the Live Template in the editor. It will auto-complete to easily add it to your code. Figure 4-34 shows the auto-complete drop-down for the `fori` Live Template.

FIGURE 4-34: Auto complete for a Live Template

After you add the `fori` Live Template to your Java code, it will highlight the variables you need to edit to implement your own custom `for` loop, as shown in Figure 4-35. When you type your own variable name and press Enter or Tab, the cursor will automatically move to the next variable to edit and so on. This action makes it very easy to customize a Live Template.

FIGURE 4-35: Cursor highlight in a Live Template

This is not the only way to add Live Templates to your code. You can also navigate to code from the Android Studio menu and click Insert Live Template or just use the shortcut (Command+J in Mac or Ctrl+J in Windows) to list all available Live Templates for the current context, as shown in Figure 4-36.

```
}   fbc                          findViewById with cast
    fixme                               adds //FIXME
    foreach                      Create a for each loop
    fori                        Create iteration loop
    gone                          Set view visibility to GONE
    I              Iterate Iterable | Array in J2SDK 5.0 syntax
    ifn                       Inserts ''if null'' statement
    inn                   Inserts ''if not null'' statement
    inst      Checks object type with instanceof and down-casts it
    IntentView            Creates an Intent with ACTION_VIEW
```

FIGURE 4-36: Insert Live Template

Another way to add a Live Template is to surround your code selection with the Live Template. After you select the code, navigate to the Code menu and then select Surround with Live Template or just press Option+Command+J on Mac Alt+Ctrl+J on other OSes to surround the selected code block with the code snippet, as shown in Figure 4-37.

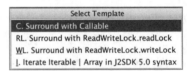

FIGURE 4-37: Surrounding Live Templates

We discussed Live Templates Java code in this section but they are available for all the structured files editable on Android Studio. If you navigate to HTML/XML, you can see the templates for HTML and XML files. If you expand the Android group, you will see around 18 (this number can change for the new versions of Android Studio) preconfigured live templates on the list. Toast Live Template is a good example for that. You can type **Toast** into your activity class and create a `Toast` object easily with Live Template.

We suggested expanding all the groups and exploring all the predefined templates. Try to use them in your application to get in the habit of using the templates. Keep in mind that there are pre-implemented code templates to launch activities, such as the starter template. If you need to define a constant (final) Java variable, just typing **const**, it will get the const Live Template and write a final variable for you.

Creating Live Templates

According to your applications, habits, or code practices you may need additional templates to speed your application development. In that case, you need to create your own template to share with your team to increase code reusability.

Creating a Live Template

In this section you create basic Live Template. In the first example, you create a function template to get the absolute value of a given variable.

Before starting to create a new template, we should navigate back to the Editor tab of Live Templates section of Preferences window (refer to Figure 4-33).

1. Select a group from the Live Template group list, click the + button and click Live Template as shown in Figure 4-38 to start adding a live template.

FIGURE 4-38: Add template button

2. Select other to add our absolute value template.

   When you select a Live Template, you will see the window shown in Figure 4-39, where you customize the Live Template.

3. Type an abbreviation, and make it short and easy to remember. We named our Live Template abs. Write the description.

   Listing 4-1 shows the necessary code to create the template. Type this code in the Template text area shown in Figure 4-39.

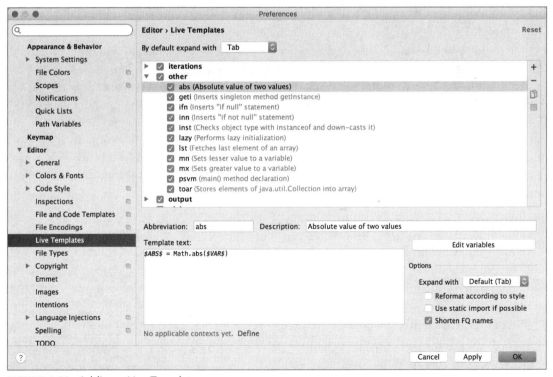

FIGURE 4-39: Adding a Live Template

LISTING 4-1: Live Template code for absolute value

```
$ABS$ = Math.abs($VAR$)
```

Variables are defined as shown in Listing 4-1, starting and ending with the $ sign. You can also configure variables by clicking the Edit variables button, which opens the window shown in Figure 4-40.

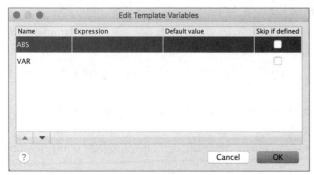

FIGURE 4-40: Edit Template Variables window

> **NOTE** *Variables can be anything that fits the context of the template. For example, an Android object can be a variable for a template in the Android group of templates.*

4. Select the context you would like to apply this template. You will see a blue text link named Define, as shown in Figure 4-39. After you click it, it will prompt the possible contexts to select from, as shown in Figure 4-41.

In this example, we selected Java because it uses Java's Math library. As you can see in Figure 4-41, you are able to select the type of text: Comment, String, Expression. and so on.

Create a Surrounding Template

The Live Templates that are able to surround selected text appear under the surround group, as shown in Figure 4-42.

Let's create a template to surround selected code text with an `if` statement that checks whether a variable is null.

The process is the same as adding a template but you also need to define where in the template the selected text will be placed. You do that by defining the `$SELECTION$` variable. You can see the sample template IFS, shown in Figure 4-42.

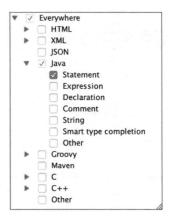

FIGURE 4-41: Live Template context list

You can also add live templates to surround selected code with block comments. If it is easier for you to remember Live Template shortcuts than the Android Studio block comment key combination (Option + Command + /), it might be easier for you to create a live template with a synonym to add a block comment.

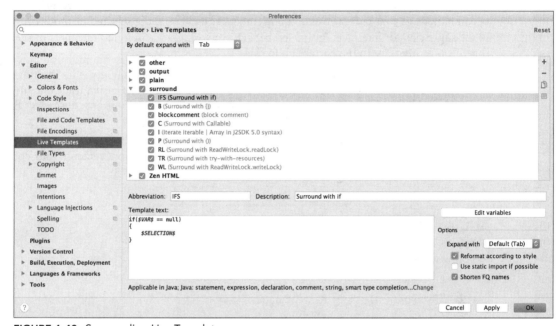

FIGURE 4-42: Surrounding Live Template

CODE REFACTORING IN ANDROID STUDIO

We visited the refactoring tools briefly in the "Android Studio Menu Items" section. This section dives into the details of what you can do with the code refactoring tools.

Having efficient refactoring tools can be a lifesaver in many situations. For example, you may change your mind a lot during development, or there may be a change in the software design, or you may want to change your naming conventions or function signatures or variable naming, and you want to change everything all at once. For these situations, Android Studio provides tools and GUI helpers to identify what you are doing.

You can see refactoring options in Figures 4-14 and 4-15, which appeared earlier in this chapter, after pressing Control+T on Mac or Ctrl+Alt+Shift+T in Windows, or you can also access refactoring options when you right-click on a Java file, as shown in Figure 4-43. The easiest method is to use the shortcut Control+T to see all the refactoring options.

Some popular refactoring options are discussed in the following list:

➤ **Rename**—You may need to rename a file, class name, or variable name. If you're changing a file or class name, you will need to look for all references to it and rename them as well. This is a mainstream tool almost all IDEs provide.

To do this, go to your code file and select a variable or any method, field, or even class name; right-click it or press Shift+F6 or Control+T (Ctrl + Alt + Shift + T for Windows/ Linux) to select the rename option. This will highlight the selection and when you finish renaming, all related references will be updated.

➤ **Pull/Push Members Up/Down**—This option helps you manage inheritance of recently developed methods in your current class. Pull Members Up will send the selected methods up to the super class or interface. Push Members Down will take the selected members of the superclass or interface to the current child class.

To use the Push Members Down or Pull Members Up option, your current class should have a base class or implement an interface. To access these methods easily, use Control+T (Ctrl + Alt + Shift + T for Windows/Linux) again to get the refactor options, and select Push Members Down or Pull Members Up to implement the interfaces.

When you select the Push Members Down option, the window shown in Figure 4-44 opens so you can select the members to push. After you've made the selection, refactor your code by pressing the Refactor button.

FIGURE 4-43: Refactoring options right-click menu

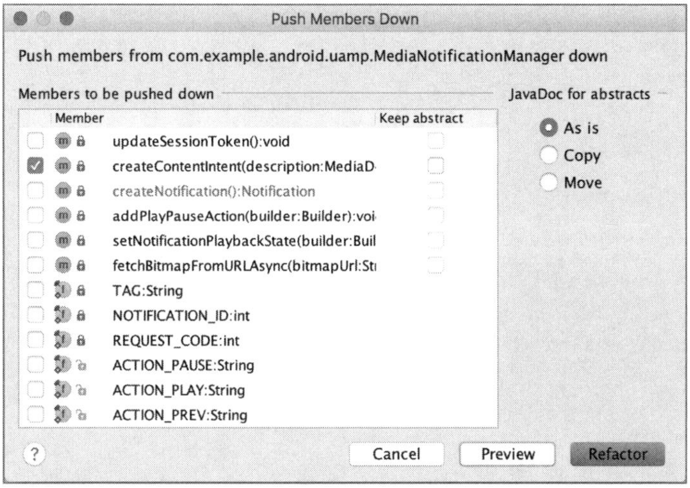

FIGURE 4-44: Push Members Down window

The Pull Members Up option works like the Pull Members Up operation, as shown in Figure 4-45.

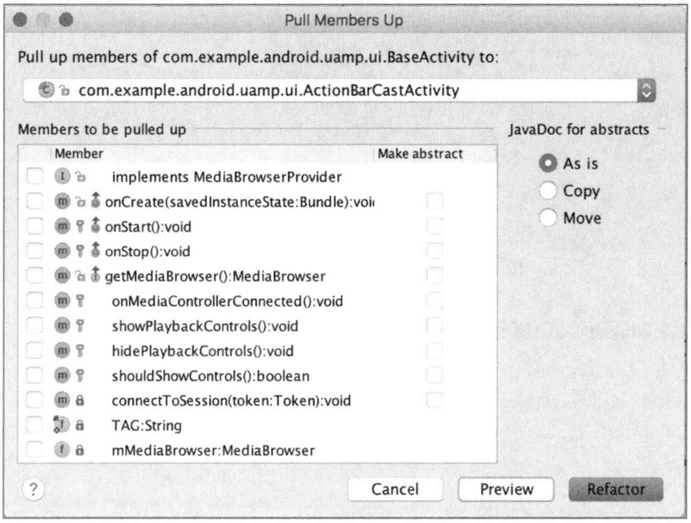

FIGURE 4-45: Pull Members Up window

➤ **Encapsulation**—This option opens a window that allows you to see and select from all the fields. Selected fields are encapsulated with getter and setter methods, as shown in Figure 4-46.

➤ **Change Signature**—This refactoring option allows you to refactor a signature name. Press Control+T and select Change Signature to see the Change Signature window shown in Figure 4-47. This window presents the visibility, return type, and name of the function. It also allows you to add and remove parameters for the existing function. It is a practical way to edit the function and complete the refactoring.

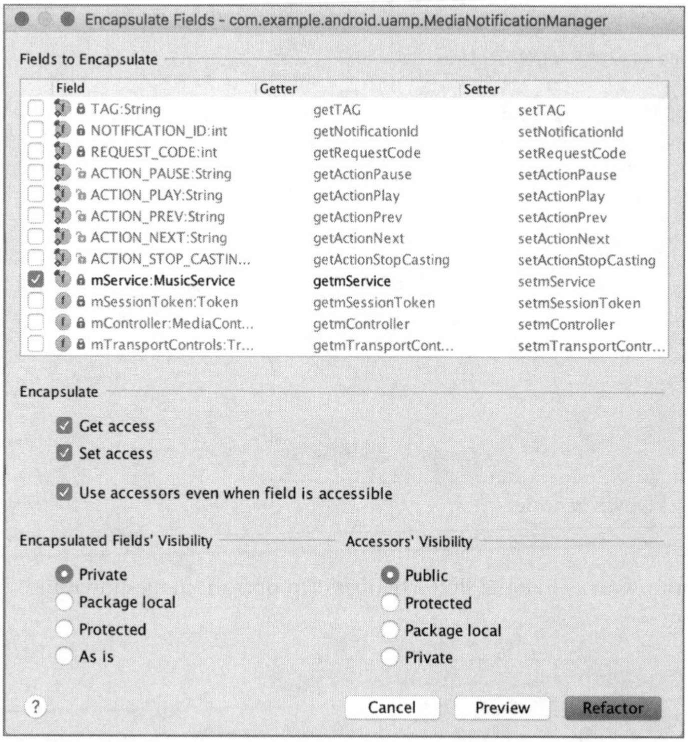

FIGURE 4-46: Encapsulation

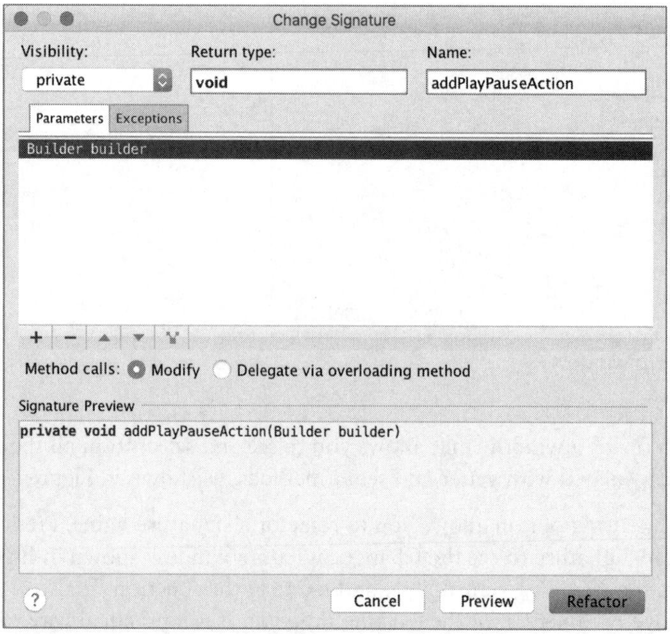

FIGURE 4-47: Change Signature window

Refactoring is a very useful technique to clean up the code you have written in a rush or while just trying out code snippets.

Another very useful concept is extraction of resources. The following list discusses the options:

➤ **Extract Variable**—Extracting an expression into a variable instead of recursive method calls is widely used to write clean, easy-to-understand code. To extract an expression into a variable, first select the expression and select Variable under Extract. The selection will be assigned into a new variable.

➤ **Extract Constant**—You may end up using the same type of `string` or `int` values over and over in your code. Extracting repeated constant values into Constants saves memory and offers better maintainability.

To extract into a constant, select the value or expression and select Constant under Extract. The selection will be turned into a `public static final` constant in class scope. You will also be asked if you want to Replace all occurrences and Move to another class.

➤ **Extract Method**—As you progress in a project, sooner or later you will notice repeating code blocks. This repetition causes the same bugs to appear in many places in a project and creates hard-to-maintain code. Extracting a code block into a method is a powerful technique to organize your code.

To do this, select the code block you want to extract and then select Method from Extract.

Next, you are asked for the visibility, method name, and parameters, followed by a preview to show your proposed method signature.

➤ **Extract Interface**—Although the extraction options covered so far are quite powerful and simple, you may need more structural changes in your class hierarchy in your development lifecycle. To extract an Interface from your class, navigate to the class in the target and select Interface from Extract.

The Extract Interface dialog box opens to let you choose an interface name, package, and methods to be added to the interface as a member.

Plus, Android Studio offers the option to rename the original class and create a copy of the original interface with selected definitions from the original interface. When you click the Refactor button, Android Studio automatically adds references and creates the new interface, and adds references where the new interface has been implemented.

➤ **Extract Superclass**—Another powerful tool to manipulate your class hierarchy is to extract a superclass from your target class. Extracting a superclass works pretty much the same as Extracting an Interface.

Navigate to the class in the target and select Superclass from Extract. A dialog box will open to let you choose a class name, package, and methods to be added to the superclass. Next to each method is a checkbox to declare the extracted method as abstract.

Android Studio offers the option *Rename original class and use superclass where possible* to change references for the given class to the newly created parent class. Clicking the Refactor button will extract the superclass.

CREATING A SIGNING KEY FOR ANDROID APPLICATIONS IN ANDROID STUDIO

Creating a signing key is an essential part of the release process. Apps built with Android Studio need to be signed to run on a device. During the development process, Android Studio uses a debug certificate, which makes the signing process almost seamless. However, signing becomes more important when it is time to release your APK to the Google Play Store. You already know that the application ID is the unique identifier. However, you need a proper way to keep the application ID safe.

An Android signing certificate is a standard key store certificate. Each key store can have more than one key and certificate.

> **NOTE** *Signing certificates can also be created by external tools. In fact, in early versions of the Eclipse-based Android Developer Tools, developers needed to create their certificates via the command line.*

To create a new signing certificate, select Create signed APK from the Build menu. Android Studio will ask you to select the module to sign, as shown in Figure 4-48.

FIGURE 4-48: Select the module to create a signed APK

Click the Next button to open the Generate Signed APK window shown in Figure 4-49. Enter the path to save your key store somewhere safe, and provide a password for the key store.

> **NOTE** *The location of your key store file is crucial. If you happen to lose your key, there is no way you can upload an update for your application. Always keep your key store file secure and make sure you don't lose it.*

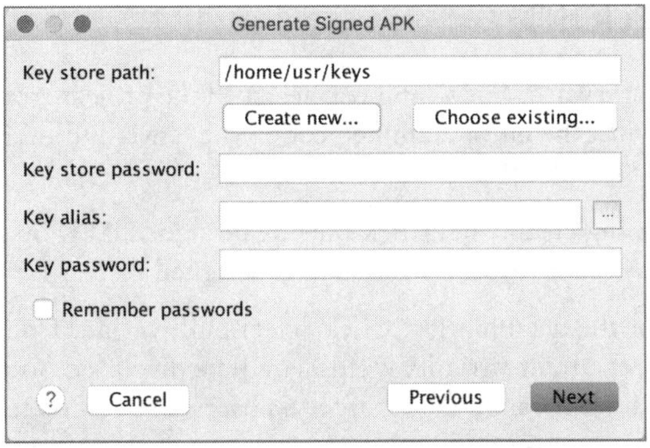

FIGURE 4-49: Create new key for the signing certificate

Next, you need to provide an alias and a password for your key and complete the rest of the fields shown in Figure 4-50. Click OK and you're done.

FIGURE 4-50: Certificate form

WARNING *Make sure not to forget your passwords. You will need to enter them each time you need to sign an APK.*

BUILDING APKS IN ANDROID STUDIO

Building APKs in Android Studio is very straightforward. Each time you select Make project from the Build menu, a debuggable APK is packaged with the latest compiled code. The newly built APK can be found under your module's `build/outputs/apk` folder.

The previous section covered how to create a signing certificate. Click the Create signed APK option from the Build menu and select your certificate. Android Studio will package a signed APK.

Android Studio lets you change build variants for the default APK. Click Select built variant from the Build menu. On the bottom right of the IDE, the built variants window will be displayed. You can choose between the debug and release build for each module. Chapter 6 covers how to create custom build variants, which can be configured from this window.

SUMMARY

This chapter touched on almost all the visible parts of Android Studio so that you are familiar with them when you need any feature in the IDE.

The chapter has offered practical advice about the functions of Android Studio for file operations, debugging, text, and code to refactor, edit, get help and access any part of the IDE for further configuration.

Our intention is that this chapter provides a useful guide to all important features: editing texts, creating templates, editing shortcuts, accessing required extra features of Android Studio to be a good Android Studio user and become an efficient Android developer.

5

Layouts with Android Studio

So far in this book, we have covered Android Studio tools, mostly dealing with files and folders and assistants to the general application development process. This chapter explores details about another important part of the application development process: *user interfaces*.

User interfaces are defined with XML files in Android applications. User interface elements in Android are configured using XML tags and attributes. The Android operating system renders the user interface using the XML file and Java code to draw the user interface elements to Android devices.

Common terms used in Android UI development are *layout*, *view*, and *widgets*. We reference these many times in this chapter and in later chapters as well.

`View` is the base class of user interface elements. Widgets are derived from the `View` class and they become a ready-to-use user interface element such as images, texts, text inputs, drop-down boxes, and so on. Layouts are the containers for views and widgets; they are group of views.

You should now create a new project, as you did in Chapter 3. Name it ChapterFive and add a BlankActivity. You will use this new project in the following section.

LAYOUTS WITH ANDROID STUDIO

Layouts define the structure for a user interface in Android applications and app widgets. In Android applications, each Android activity handles the user interface operations so all activities have a layout design. App widgets are the small interaction interfaces placed in the home screen of Android for easy access to main functions of an application. Layouts are being used for the structured design of these interfaces.

XML is used to create a layout, which is stored in the res /layout folder. The build system automatically recognizes XML files in the layout folder as a user interface element to be displayed.

When you create a new application, you select an activity, for which layouts are auto-generated and placed in the layout folder. If you created the ChapterFive project with a BlankActivity, as suggested in the chapter introduction, you will see the layouts shown in Figure 5-1.

Figure 5-1 shows two layout files and one activity, which means that you can use layouts inside layouts to create reusable user interface structures in multiple activities and ease the refactoring of the user interface design.

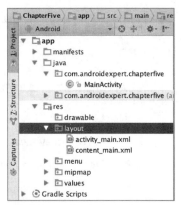

FIGURE 5-1: Layouts in the Project view

Adding a New Layout File

Now that you know where layouts are stored, let's add a new layout to your recently created application. In Android Studio, it is pretty easy to identify a layout file and add it to your project. Go to File ➪ New, select XML at the bottom of the menu, and then Layout XML, as shown in Figure 5-2.

After you click Layout XML File, the window shown in Figure 5-3 opens. In this window, you can name the layout and select the Root Tag.

The Root Tag field shows the structure type for the layout. The following list describes layout tags you can use to arrange views.

➤ **Linear Layout**—Linear Layout is used to arrange views as a single row with multiple columns or a single column with multiple rows. Each element comes either after or beside the previous element. Designers can choose to arrange this type of layout as they prefer. This is an easy layout to use.

➤ **Relative Layout**—Relative Layout is used to arrange views relative to another view's position. This layout is a little harder to manage than the Linear Layout because it is the designer's task to position elements relative to one another.

➤ **Frame Layout**—Frame Layout is used as a placeholder layout for a single view. It makes it easier to arrange a view's position in another layout. Video playback is a good Frame Layout use case. Because video playback is done using a single video view, assigning a video view to a Frame Layout would make video seem better aligned with the application user interface.

➤ **Table Layout**—Table Layout creates table-like views with columns and rows. You can choose how many columns you want for each row.

➤ **Grid Layout**—Grid Layout offers the ability to create a user interface with elements arranged in multiple rows and columns. For example, if you need to list multiple photos in the same screen at once, you can create a grid to show their thumbnails easily. Although Grid Layout offers functionality that's similar to Table Layout, it is more like a mixture of Table Layout and Linear Layout.

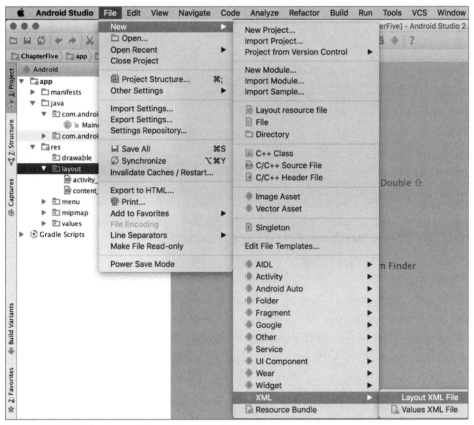

FIGURE 5-2: Adding a new layout in Android Studio

> **NOTE** *In addition to the preceding list of layouts, List/Recycler/GridView are useful views that display items in their own layouts using adapters, and should be used to display dynamic data efficiently. However, they are not used as a layout in Android SDK, so we don't cover them in this chapter.*

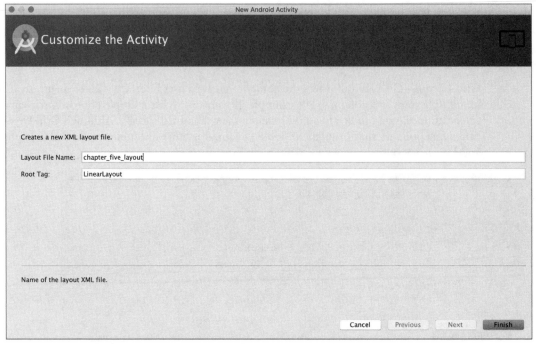

FIGURE 5-3: Configuring a new layout file

Layout Design Structure in Android Studio

Now, let's discover Android Studio's basic layout design and development properties. We go into greater detail in later sections.

Right after you add a new XML layout, Android Studio opens it as an XML file or in Design mode. If a layout has been opened in Text mode, you will see the content of the XML file: either just the file, or the layout preview as well, as shown in Figure 5-4.

In case the preview is not opened by default, you can activate it with the Preview button on the edge of Android Studio or you can navigate to the Tool Windows option on the View menu and select Preview.

> **NOTE** *Preview is a recent enhancement in Android Studio, as compared to Eclipse IDE. In Eclipse, you would switch to the Design perspective to see the effects of changes made in a layout XML file. Now, with Android Studio, you are able to directly see the preview of the layout right after you make changes to the XML file.*

For a professional Android UI developer, using XML files is the best way to develop the user interfaces. However, if you want to drag and drop, Android Studio provides the Palette tool shown in Figure 5-5, with layouts and views to easily design the user interface. To switch between design perspectives, you can use the Text and Design buttons.

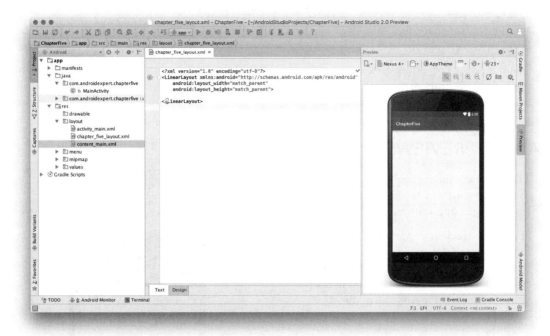

FIGURE 5-4: Layout in text mode

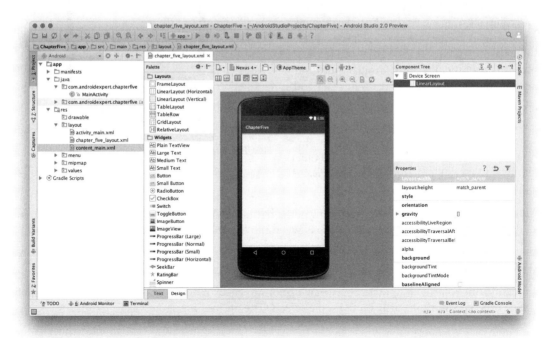

FIGURE 5-5: Visually designing in Android Studio

The Palette view shows a preview of the layout for the selected device. In Design view, you see a component tree that defines the child and parent relationships with additional views in the layout. You can also change the order of views in the component tree because some layouts require that.

There is also a Properties window where you can edit the properties of a layout, view, and all other user interface components. In Text mode, the editing is done within the XML tags. We revisit the design details with Android Studio in later chapters.

LAYOUT PREVIEWS

The Preview window helps designers see the changes applied to a layout instantly without launching the application.

Having an efficient preview tool is really important for GUI-based application development. This is especially true for mobile applications, which rely on a good user experience, especially for Android application development.

Android is growing faster than any mobile platform, not only on mobile phones but on wearables, TVs, cars, and Google Glass. There are probably hundreds of different screen sizes running Android, so developing a generic user interface or developing a user interface for the most used sizes and resolutions is really critical in Android application development to be sure all users have a good experience.

With the Preview tool, you can test your layout on any kind of screen for defined resolutions, the orientation of the device, and older

FIGURE 5-6: Preview window

Android SDKs. These functions are accessed by pressing the appropriate buttons, some of which are shown in Figure 5-6.

For example, clicking the third button from the left in Figure 5-6 rotates the device image horizontally so you can see how your layout looks when the target device's orientation changes.

Layout Rendering Options

Android Studio provides practical preview options to easily render the layout for all devices and SDK versions simultaneously on the same screen. The options are shown in Figure 5-7.

Having rendering options that enable you to see preview-based Android versions and screen sizes eliminates the work of opening each rendering option one by one, which makes a UI designer's life easier.

FIGURE 5-7: Layout rendering options

Previewing Virtual Device Views

Because Android is an open source operating system, many companies, including large global brands and local OEMs, have customized and used it on their own hardware devices. As a result, you can see Android running on screens of almost any size.

Testing applications on the many common devices used by consumers is a necessary but painful and slow process. Launching applications on a wide variety of virtual devices during development is an option, but it is a very time-consuming process. Another option is to try layouts on real devices, but this is also a time-consuming and expensive process.

Android Studio improves this process by supplying predefined screen devices for previewing the layout during development. You can select devices by clicking the button with a device name to display the list of available previews, as shown in Figure 5-8.

You can also select tablet and TV-sized devices, and wear devices. As Figure 5-8 shows, entries in the list match the name, screen size, and resolution of the target product. If you want to test with other, generic devices, navigate to the Generic Phones and Tablets option to see the list shown in Figure 5-9.

Figure 5-9 shows the list of *predefined* generic screen resolutions and sizes for testing. If you want to define a *new* size or resolution you can select the Add Device Definition option, which leads you through the process of creating a new virtual device. Once created, you can use its settings in the preview view. As you saw in Figure 5-8, we use a virtual device for previews (the Nexus 4 option that is checked).

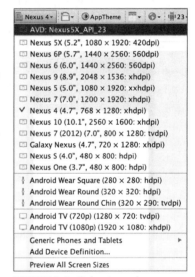

FIGURE 5-8: Selecting a layout to preview in Android Studio

Previewing on Different Android SDK Versions

Previewing your layout according to size and resolution is not enough if your application's audience is large; you should also check that the layout is rendered the same in the current Android SDK. Many people use different Android devices with different release versions, so you can't be sure that every-body will use the same version at the same time and that your UI will work the same in all of them.

At each big release of Android, there can be minor or major changes to the API, which is directly related to UI layout. For example, FrameLayout is used a lot but was not a defined layout in earlier releases of Android and so may not be rendered in older versions. To avoid a situation like this, install the Android SDKs you want to test the UI on, and preview your layout with the major SDK versions, as shown Figure 5-10.

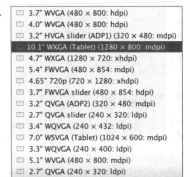

Selecting Themes

A theme is a set of styles applied to all user interface elements to keep them looking similar in all application windows. There are many examples. A font style is a theme: When you define a font and apply it to all text views in the application, all text will be similar. A predefined color style also can be applied to all views to keep them in a similar color.

FIGURE 5-9: List of generic devices

Predefined themes for Android applications come with Android SDK. You will noticed that a predefined theme has been assigned by Android Studio while previewing your layout. However, there are additional style options for layouts. Just click the Theme Selection button (to the right of the device orientation button shown in Figure 5-6) to open the window shown in Figure 5-11.

FIGURE 5-10: Picking the Android SDK version for previewing

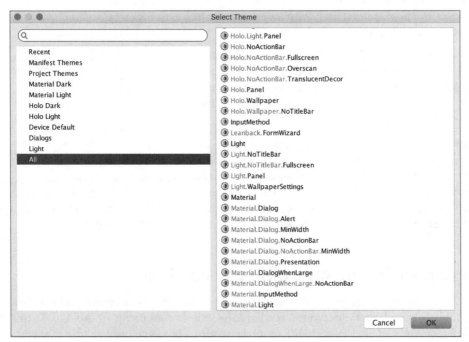

FIGURE 5-11: Theme selection window

Themes enable you to observe the appearance of the layout in any selected application style. Using themes makes styling recurring views with same look easier. Views such as dialogs and alerts can show different content but have the same appearance by applying same theme to those UI elements. Developers might want their application to be either full screen or only show the Android action bar while running; this can also be done using themes. (Notice the themes ending with **NoActionBar** in Figure 5-11.)

DESIGNING LAYOUTS WITH ANDROID STUDIO

In this section, you add user interface components and views to the layout you are designing.

> **NOTE** *Layouts can also be designed within Java code, but it is not a recommended development practice.*

There are two ways to add views to your layout. The first option is to use the Palette tool to add a predefined view or layout to your user layout. Figure 5-12 shows part of the Pallet.

FIGURE 5-12: Palette tool

> **NOTE** *Layouts can contain one or more other layouts. This way you can reuse a layout you previously designed in many other layouts.*

The Palette tool includes the following views and layouts:

➤ **Widgets**—Widgets are single views that present a user element such as TextView, RadioButton, CheckBox, Button, and Switch.

➤ **Text Fields**—Contains predefined text fields for different types of text inputs such as password, date time, number, and e-mail. Selecting a text field with a predefined input type will make keyboard input easier. For example, when you select a number field, the keyboard opens with only numbers available.

➤ **Containers**—This group of views can contain any view type such as RadioGroup, ListView, GridView, SearchView, and VideoView.

➤ **Date Time**—This group includes date- and time-related views.

➤ **Expert**—This group includes view items with advanced uses such as the TextureView and SurfaceView, which are used to render OpenGL graphics. They are used primarily by game developers. Resources that use OpenGL will be presented using the SurfaceView.

➤ **Custom**—This group has four items to define customized views that are not included in the palette.

Let's play with a Plain TextView widget in our layout.

1. First drag a Plain TextView widget (refer to Figure 5-12) from the pallet to the preview.

The Plain TextView widget will show up in the layout preview, and an XML entry is created in the Text mode view. According to your choice of layout, its location on the screen may change. Figure 5-13 shows that views are added vertically. If LinearLayout was chosen to be horizontal, views would be added side by side, horizontally.

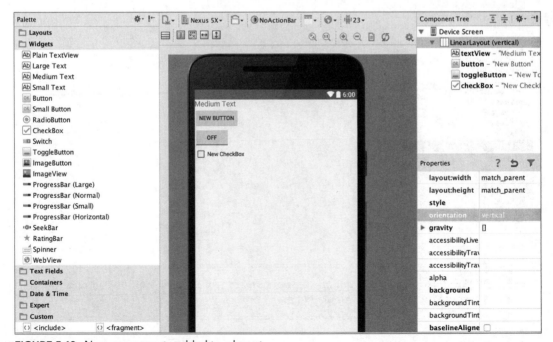

FIGURE 5-13: New components added to a layout

2. After adding the views, you may want to make changes to the view's ID, text, position, background color, and position. You can do this from the Properties pane at the right of the Designer window shown in Figure 5-13. Setting the `id` property gives a unique ID for the view. If you set the `text` property to Chapter Five, the text Medium Text will disappear and you will see Chapter Five in the preview window.

3. In the Properties pane, you can change the text's width by setting the `layout:width` property to `fill_parent` to extend it and use all the available vertical space.

4. Finally, set the text's position to center by setting the `gravity` property to `center_horizontal`. When you set the view's property to center, the text will be centered. You can also set it to `left` or `right`, which will move the text in that direction.

    After you change the `width` and `gravity` properties, you see that the view has been changed as well, as in Figure 5-14.

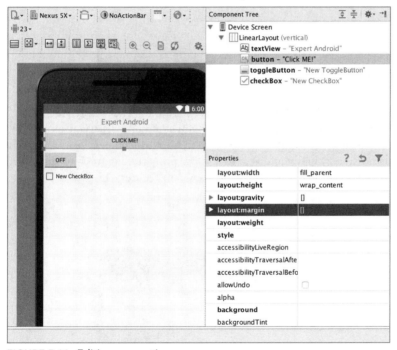

FIGURE 5-14: Editing properties

In the Properties window, some items have a list of values to set and some are set manually. You can also access the Help documentation by clicking the question mark (?) button on the top right. Clicking the rightmost button reveals the hidden properties for more advanced view settings.

To set properties in XML format, you have to know the names of the elements because you have to type them. However, Android Studio provides help by listing all the options right after you type < to start a new tag in the XML file, as shown in Figure 5-15.

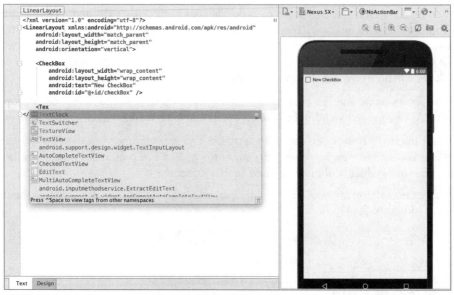

FIGURE 5-15: Adding a new component in XML files

After adding the element tag in the XML file, you can edit the attributes. You will see the possible properties listed when you hit Control + spacebar. After you add a property, you set its value by first typing two double quote marks (**""**). Then place the cursor between them and press Control+spacebar to list the possible values you can assign to the property. Figure 5-16 shows an example of changing the properties of the CheckBox View shown in Figure 5-15.

FIGURE 5-16: Editing view properties in XML

MANAGING RESOURCES

To make a better Android UI, you need to know how to use static resources, strings, color definitions, style definitions, dimensional definitions, and drawables.

Resources are the main building blocks for layouts. They are reusable and easy to manage, and they enable developers to create consistency over all their user interfaces. Relying on the Android guidelines helps make your application easy for end users.

This section provides a brief overview of strings, styles, colors, dimens, and drawable resources and provides examples of how to use them in the layouts.

Using Strings

While developing user interfaces, you will need to use text on the screen. The Android guidelines suggest adding text to user elements using string resources instead of just writing the text to XML files or to a text property.

You create string resources in another XML file containing a list of elements. Initially there are only two items, but for this example we added new ones in a similar style—using the string tag in the XML file. Listing 5-1 shows the content of the `string.xml` file located in the `res/values` folder.

LISTING 5-1: string.xml content, string resources

```
<resources>
    <string name="app_name">ChapterFive</string>
    <string name="action_settings">Settings</string>
    <string name="ok_button">OK</string>
    <string name="chapter_name">ChapterFive</string>
    <string name="next">Next</string>
</resources>
```

To use a string resource in the UI layout, you assign a property's text starting with `@string` in the XML file. For example:

```
"android:text="@string/chapter_name"
```

You can also do this using the Properties window. With the mouse pointer on the `property` field, click the button that appears and a window will pop up that lists all the resources you can use, as shown in Figure 5-17. Select the string resource to assign to your view.

Notice the New Resource button at the bottom of Figure 5-17. It is another alternative to add a new string resource to the `string.xml` file.

Using Styles

String resources are not the only resources you can reuse for your views. There are also style resources to assign to your view to change the appearance of text, including background color, size of text, font, shadowing, and so on.

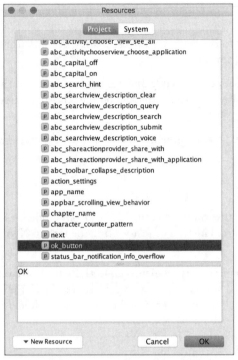

FIGURE 5-17: Resource selection window

Style resources are stored in the `res/values/styles.xml` file. Some predefined styles are already in the `styles.xml` file. Let's quickly make a new style to apply our layouts.

Listing 5-2 shows a simple style resource in the `styles.xml` file. It uses the `style` XML tag to create a new style to change the text color, size, padding, gravity (placement on the layout), and font.

LISTING 5-2: Style sample

```
<style name="NewStyle" parent="AppTheme">
    <item name="android:textColor">#229933</item>
    <item name="android:textSize">20sp</item>
    <item name="android:padding">5sp</item>
    <item name="android:typeface">monospace</item>
    <item name="android:gravity" >center</item>
</style>
```

To apply a style to a view, in the text mode of layout, add the `style="@style/NewStyle"` line or find the `style` property to select the NewStyle resource from the list. After you apply the style to the view, you will see that the view element changed in the preview. See Listing 5-4 in the following section for a sample style assignment in an XML layout file.

Styles are important. To make your user interface consistent, reusing and applying styles to all views will make them seem more professional and ordered.

Using Dimens

Dimen definitions are used to change the width, length, height, and margins of the assigned view in the layout for devices with different resolutions and screen sizes.

Using dimens gives you the flexibility to both design your layout for a specific device size and to assign a `dimen` resource to change to a different device size at runtime. You can create a better experience with all devices using the `dimen` resource.

Dimen tags are stored in the `res/values/dimens.xml` file. Dimension resources are created under the `resources` XML tag with `dimen` tags, as shown in Listing 5-3, which is the `dimen.xml` file generated right after the project was created.

LISTING 5-3: dimen.xml content

```
<resources>
    <!-- Default screen margins, per the Android Design guidelines. -->
    <dimen name="activity_horizontal_margin">16dp</dimen>
    <dimen name="activity_vertical_margin">16dp</dimen>
    <dimen name="fab_margin">16dp</dimen>
</resources>
```

To use the `dimen` resource right after definition, you need to find the resource that is related with the dimensions of the view. For example you can go to the `layout:margin` property and click the button to select from the available resources in the window that opens.

Assignment of the `dimen` resource in XML files is similar to assignment for strings and styles. For example if you want to assign `fab_margin` to your check box, added in previous section, you need to define the `layout:margin` attribute and assign the `dimen` with `"@dimen/fab_margin"`, as in Listing 5-4.

LISTING 5-4: Sample view with style, dimen, and string resources

```
<CheckBox
    android:layout_width="match_parent"
    android:layout_height="wrap_content"
    android:gravity="center"
    android:text="@string/chapter_name"
    android:id="@+id/checkBox"
    style="@style/NewStyle"
    android:layout_margin="@dimen/fab_margin" />
```

All the defined resources are auto-referenced by Android Studio, so right after you start typing `@dimen`, `@string`, and `@style` recently defined resources will be listed.

Using Colors

Color definition resources can be used to assign layout and view color-related properties such as background, text color, border color, and so on. Color resources are defined in the res/values/colors.xml file. Right after you define the color, the editor in Android Studio provides a preview of the color on the left (a colored square), as shown in Figure 5-18.

FIGURE 5-18: Color resources definition

As you do for other resources, you assign a color by finding the property that can take a color (such as background) and assign the resource.

In an XML file, typing android:background="@color/colorBlack" makes the background black.

Using Drawables

Drawables are resources which contain graphics that can be drawn on screen. They can be applied directly in the XML layout file with the android:drawable attribute. As in the previous example, you can add a color value as a drawable with android:drawable="@color/colorBlack".

Of course, drawables are not limited to color values and can be used for bitmaps and nine patch files; state, level, and layer lists; and transition, clip, scale, shape, and inset drawables.

Drawables can be defined in XML format. Listing 5-5 provides a simple example to draw a rectangle with rounded corners and has a gradient fill.

LISTING 5-5: Drawable definition

```
<shape
  xmlns:android="http://schemas.android.com/apk/res/android"
  android:shape="rectangle">
<corners
    android:bottomRightRadius="10dp"
    android:bottomLeftRadius="10dp"
    android:topLeftRadius="10dp"
    android:topRightRadius="10dp" />
<gradient
    android:endColor="#DD888888"
    android:startColor="#DD333333"
    android:angle="45" />
</shape>
```

Other drawables that can be declared in XML are vector, transition, state, and animation drawables.

USING LAYOUT TOOLS

Android Studio provides developers translation capabiltiy to ease localization of Android UI text, and Activity Association to assign Android Activities to layouts and indicate the application themes on previews.

Translation

As a developer, you might be surprised that adding a translation will boost your downloads. There are several markets and languages that have a large Android ecosystem but low levels of English adaptation.

Translating your text resources can be simpler than you might imagine if you have externalized all the strings. As you saw in the "Using Strings" section, each string value should be externalized in the `strings.xml` file and never under any condition hard coded.

Right-click the `strings.xml` file and select Open Translations Editor. This editor not only serves as a key/value strings editor but also helps you to localize your resources. Figure 5-19 shows the Translation Editor.

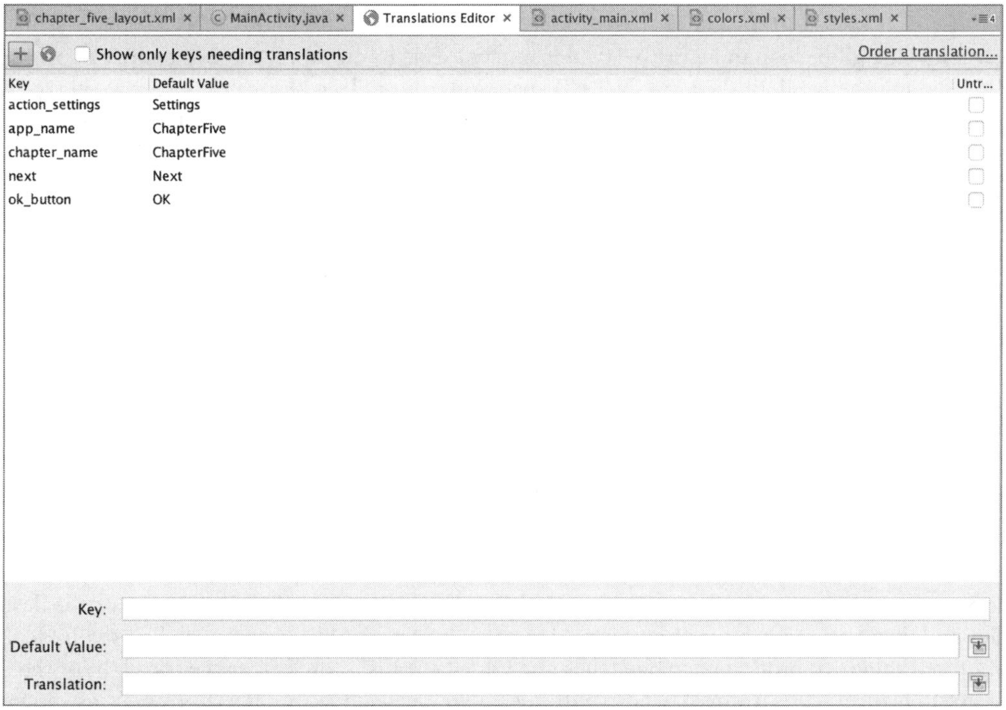

FIGURE 5-19: Translation Editor

The green plus sign icon at the top left of the window creates a new key value pair, as shown in Figure 5-19. Because you don't have any other locale yet, the created values are the default locale for the app. The default locale acts as the fallback locale if no localization matches the user locale. Figure 5-20 shows the expanded list of locals.

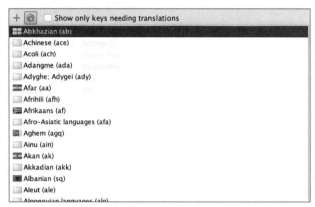

FIGURE 5-20: Expanded list of locales

The second blue globe icon in Figure 5-10 adds a new locale to our app. You can choose to add several locales and they will all show up in the Translation Editor. To edit a value, select a cell from any locale and the key, default value, and the translated value you wish to enter will be displayed at the bottom of the editor, as shown in Figure 5-21.

Key:	greetings
Default Value:	Welcome
Translation:	Bienvenue

FIGURE 5-21: Translated and default value

Ideally an application should have a full list of values for each key in each locale. If a value is missing, the default for that value will be used instead. To check for missing values, click the Show only keys needing translations check box and the editor will filter out the complete localized keys.

Supporting many locales looks pretty easy but in practice can be very troublesome. How would you support an East Asian language with a different alphabet if you do not have any knowledge of the target language? Fortunately, Google has a solution for that. You can buy a language for the desired locales and then you can ask for only the needed value using the filter tool, which helps you identify the required value. To order a language, click the Order a translation link at the top right of the editor (refer to Figure 5-19). Android Studio will open a web page in your default browser, as shown in Figure 5-22.

Next, you need to select your source language and upload the XML file. In the next step, you will be asked for the target languages you want to order. This is a paid service and you pay for the number of languages and items you need, but it's well worth the expense because with the right set of locales, your market accessibility will dramatically increase.

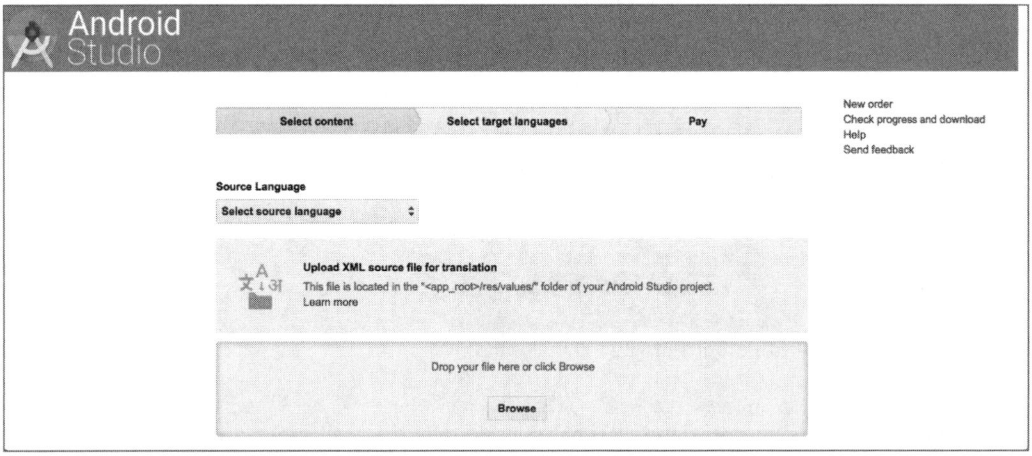

FIGURE 5-22: Android Studio language ordering web page

Activity Association

All layouts are associated with an activity during runtime. However, because this association comes during runtime, it can be difficult to see a preview of the layout during development because you might miss the theme selection for the activity, which is made in the `AndroidManifest.xml` file with a setting similar to the following:

```
android:theme="@style/AppTheme.NoActionBar"
```

There are two ways to set the layout's context: either associate an activity visually in Design view or in the XML file. In the XML file, you need to associate an activity using the `tools:context` attribute in the layout tag itself, as shown in Listing 5-6.

LISTING 5-6: Theme association in layout

```
<?xml version="1.0" encoding="utf-8"?>
<LinearLayout xmlns:android="http://schemas.android.com/apk/res/android"
    xmlns:tools="http://schemas.android.com/tools"
    android:layout_width="match_parent"
    android:layout_height="match_parent"
    android:orientation="vertical"
    android:weightSum="1"
    android:background="@color/colorBlack"
    tools:context=".MainActivity">

    <CheckBox
        android:layout_width="match_parent"
```

```
                android:layout_height="wrap_content"
                android:gravity="center"
                android:text="@string/chapter_name"
                android:id="@+id/checkBox"
                style="@style/NewStyle"
                android:layout_margin="@dimen/fab_margin" />

    </LinearLayout>
```

In Design view, click the MainActivity button and select the Associate with other Activity option as shown in Figure 5-23.

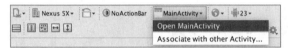

FIGURE 5-23: Associate with other Activity

After association, the layout preview will change if your Activity theme is different than the currently selected theme.

ASSET MANAGEMENT

Asset management is one of the most undervalued topics in Android development. Not surprisingly, developers like to write code and may tend to skip best practices when it comes to assets such as images and bitmaps.

Since the early versions of Android, Google invested in tools to help developers deal with assets. The draw nine patch tool was one of the early tools provided for handling images that can be displayed properly on different screen sizes. Android Asset Studio and other tools integrated with ADT followed.

Android Studio is integrated with tools and offers different project views to group and organize assets. Each Android module has a res folder to host all assets as well as other resources except for source code. Because the Android ecosystem has a wide variety of devices with different screen sizes and densities, assets that can handle these different screen sizes are needed. Figure 5-24 shows asset folders to address resolution/density differences.

Early on, Android limited this list to ldpi, mdpi, and hdpi, but with the higher density pixels and larger screens available

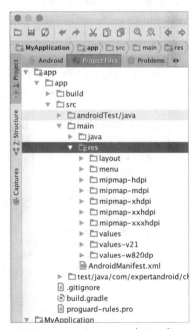

FIGURE 5-24: Asset resolution list

today, you have up to xxxhdpi density. Ideally, an image file should be resized for each screen resolution/density your app supports.

Android view (as opposed to Project view shown in Figure 5-24) offers an easier-to-follow visual structure for grouping the asset files (see Figure 5-25).

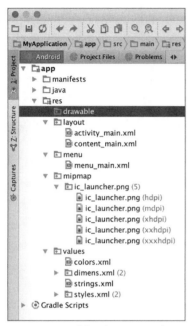

FIGURE 5-25: Visual structure for asset resources

This view groups different image sizes under the same resource name, in the mipmap folder, and also displays dpi info and the number of files in the folder. Android Studio handles all resizing processing when an image resource is created or imported via Android Studio. To import an existing image, right-click the project and select Image Asset from New, as shown in Figure 5-26.

Next, the Asset Studio window is displayed for importing your image, as shown in Figure 5-27.

Asset Studio creates different sizes of resources from your input and places them into the appropriate folders.

The wizard has several options, as follows:

➤ **Asset Type**—Allows you to choose icons between launcher, action bar/tab, or notification (see Figure 5-28).

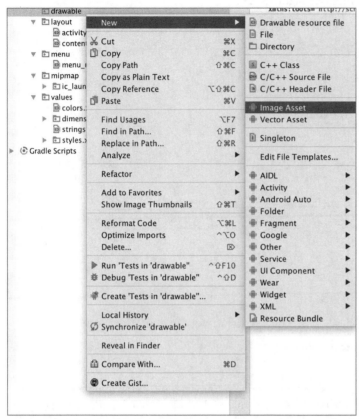

FIGURE 5-26: Adding an image asset

➤ **Foreground**—Allows you to select an image, clip art, or text as an asset (see Figure 5-29).

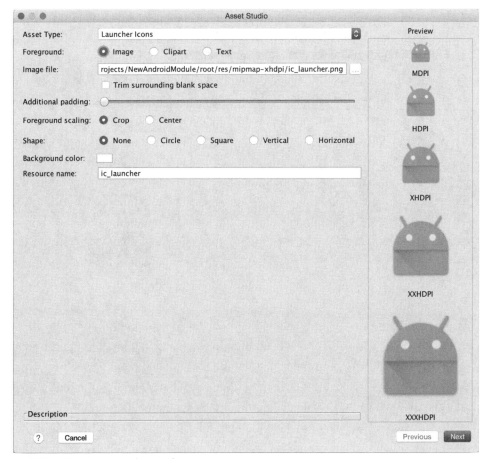

FIGURE 5-27: Asset Studio window

➤ **Image File**—Allows you to choose an image.

➤ **Additional Padding**—Increases the space around the image.

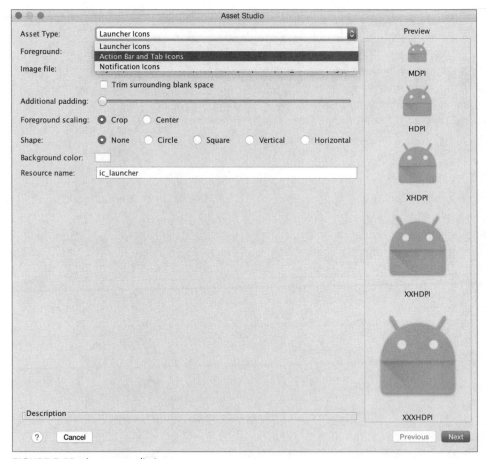

FIGURE 5-28: Asset type listing

➤ **Foreground Scaling**—Allows you to choose between cropping or centering the image during resizing.

➤ **Shape**—Allows you to choose the output image shape.

➤ **Background Color**—Allows you to choose a background color for the created images.

➤ **Resource Name**—Allows you to decide on the output file name. All the various sizes of the same image will be named the same but will be placed into appropriate folders based on the image size.

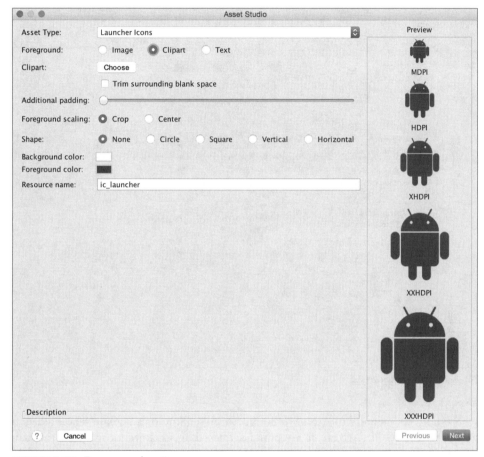

FIGURE 5-29: Foreground options

Another useful tool for creating image resources is the Vector Asset Studio. To create a vector asset, right-click on the project tree and select Vector Asset under New, as shown in Figure 5-30.

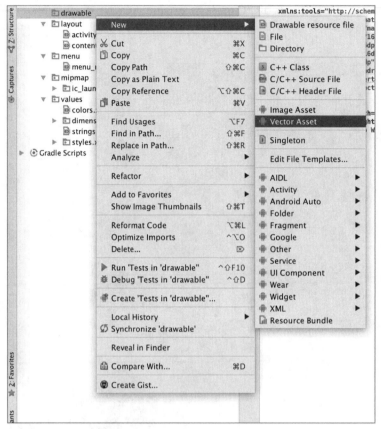

FIGURE 5-30: Adding a vector asset

Vector Asset Studio is very similar to Asset Studio. Vector Asset Studio can create vector assets from material icons or local SVG files. To create a vector asset based on a material icon, select the first option, as shown in Figure 5-31.

Next, configure vector by selecting icon, setting size, and opacity from Vector Asset Studio window:

➤ **Icon**—An icon that the vector asset will be built from

➤ **Size**—The output size of the asset in terms of dp

➤ **Opacity**—The opacity of the created vector asset

The Local SVG file option offers the same configuration except that it asks for a file instead of an icon.

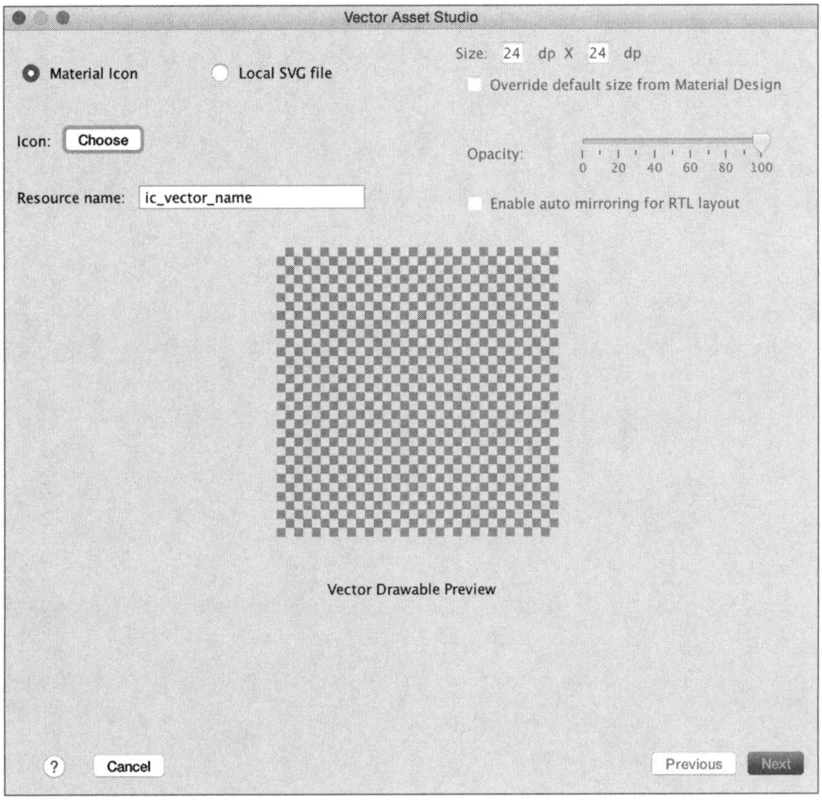

FIGURE 5-31: Creating a vector asset

SUMMARY

This chapter covered the details of user interface design and development with a focus on layout. User experience is at the heart of Android application development to reach a wider audience. We covered the functions, including resource management, that we think are important and which you will need to use repeatedly during layout development with Android Studio.

> **NOTE** *Read our blog at* http://www.devchronicles.com/2016/06/expert-android-studio-book-updates.html *to see the changes announced at Google I/O 2016.*

Android Build System

WHAT'S IN THIS CHAPTER?

➤ Android build system: Gradle

➤ Using Gradle

➤ Managing dependencies

➤ Configuring Android Plugin for Gradle

➤ Writing a Gradle plugin

Android Studio has introduced many changes to the Android development lifecycle that are limited not only to the IDE and tools but also to the build system. A Gradle-based build system was introduced with the initial release of Android Studio.

Prior to Android Studio, the Android ecosystem did not have one default build system. Some developers relied on Apache Ant scripts, whereas other developers preferred more sophisticated Maven builds. Another popular way to build Android apps uses mk files, which were widely used by developers using the Native Development Kit (NDK).

A common and yet simple approach followed by developers was to copy libraries (jar or aar files) into the libs folder and let Eclipse build tools to handle the build. However, this approach created problems when the project was integrated with source control systems. Although Maven addressed most of the dependency and automated test/build issues, it introduced another layer of complexity and performance problems.

In this chapter, you learn how to use Gradle effectively to control builds, manage dependencies and, even better, how to add custom tasks by writing your own plugins.

USING GRADLE

The Gradle build system was first released in 2007. Unlike Maven, which relies on XML, Gradle uses a Groovy-based domain-specific language for project configuration.

Basically, Gradle offers a simpler syntax to declare dependencies and build properties. It can easily be extended and used for complicated tasks and large projects. Gradle uses a directed acyclic graph to determine the order of the tasks. Gradle is widely used to build Java, Scala, and, of course, Groovy projects.

Gradle met Android in the release of Android Studio. Android Studio comes with a Gradle wrapper for seamless integration with Gradle. The Android build system offers an Android Plugin for Gradle, which not only takes care of all IDE-based compiles and builds but Gradle can also run standalone even when Android Studio is not installed. This allows Android projects to be easily integrated with Continuous Integration servers such as Hudson and Jenkins.

Anatomy of Gradle

Gradle build configuration is defined in the build.gradle files in Android projects. Build files exist both in modules and in the project to configure properties related to the given scope. A build file typically contains Android plugins to configure your project.

Project scope is defined in the build.gradle file and is mainly used to declare project-wide repositories and dependencies, as shown in Listing 6-1.

LISTING 6-1: build.gradle file content

```
buildscript {
    repositories {
        mavenCentral()
    }
    dependencies {
        classpath 'com.android.tools.build:gradle:1.3.0'
    }
}

allprojects {
    repositories {
        mavenCentral()
    }
}
```

This build file adds `mavenCentral` as a repository and the `classpath` dependency for the Android Plugin for Gradle version 1.3.

> **NOTE** *The Android Plugin for Gradle has introduced some changes that broke backward compatibility in the past. An infamous change that resulted in broken builds happened with version 1.0 when the* `runProguard` *property was changed to* `minifyEnabled`. *Always take a look at the change log before upgrading your Android Studio or Android Plugin for Gradle.*

In addition to the project-scope build.gradle file, each module has its own build.gradle file for project-specific configuration. The module-scope build file is where the Android Plugin for Gradle really kicks in and works its magic. The module build file offers the user numerous options, such as the capability to override the manifest settings and to change the app package, source, resources, and ID.

The Android Plugin for Gradle can configure the following:

➤ Android settings, such as `compileSdkVersion` and `buildToolsVersion`

➤ Product flavors and `defaultConfig`, which can override `applicationId`, `minSdkVersion`, `targetSdkVersion`, and test information

➤ Build types such as debug, version name, and ProGuard configuration

➤ Dependencies such as external, local, or other modules

Listing 6-2 shows a typical module Gradle file.

LISTING 6-2: Module Gradle content

```
buildscript {
    repositories {
        jcenter()
    }

    dependencies {
        classpath 'com.android.tools.build:gradle:1.3.0'
    }
}

apply plugin: 'com.android.application'

repositories {
    jcenter()
}

dependencies {
    compile "com.android.support:support-v4:23.1.0"
```

```
        compile "com.android.support:support-v13:23.1.0"
        compile "com.android.support:cardview-v7:23.1.0"
        compile 'com.android.support:appcompat-v7:23.0.0'
    }

android {
    compileSdkVersion 23
    buildToolsVersion "23.0.2"

    defaultConfig {
        minSdkVersion 21
        targetSdkVersion 23
    }

    compileOptions {
        sourceCompatibility JavaVersion.VERSION_1_7
        targetCompatibility JavaVersion.VERSION_1_7
    }

}
```

The first part of the build script declares repositories and dependencies for the module. As discussed previously, you can use this configuration on both the project and module scope. These dependencies are Gradle dependencies and should not be mixed with Android project dependencies. In this example, we simply add the Android plugin for Gradle version 1.3.0 to make Gradle and Android Studio work in harmony to build our Android project.

Next, you need to apply the Android Plugin for Gradle you have just added as a dependency. The `apply plugin:` task followed by the plugin name does the magic. You can also choose to apply other Gradle plugins, which would offer other tasks and functionality. This is covered in the "Writing Your Own Gradle Plugin" section later in this chapter.

Once the Android Plugin for Gradle is applied, you can declare Android dependencies for the given module. In this example, you use four support libraries from Google, which provides support to use new widgets, APIs and libraries on older versions of Android. With the help of support libraries, you can keep your `minSdk` level to target older versions while being able to use cool newly released functionality.

You are almost there; finally, you can configure the Android plugin for Gradle in the `Android` block. The Android Plugin for Gradle offers many capabilities, which we cover in this chapter.

Listing 6-2 gives a basic example that sets SDK and tool versions as well as declaring a version of Java for the compile options. Although you may not need to tweak those configurations daily, you definitely need to learn the details in order to have full control of your project. For example, Retrolambda, a popular third-party open source library that lets you use Java 8 syntax on Android, requires you to set the Java version to 8 in order for the Android Plugin for Gradle and Android Studio to function properly.

DEPENDENCY MANAGEMENT WITH GRADLE

Gradle offers a great way to handle project dependencies without the need to copy source code from project to project. Even better is that Gradle's way to declare dependencies is very simple when compared to Maven, yet still very flexible and customizable. Gradle really shines when it comes to dealing with dependencies.

Gradle offers different scopes for declaring dependencies:

➤ **Compile**—Declares dependencies that are required to compile the project from source code.

➤ **Runtime**—Declares dependencies that are needed during the execution of the compiled code. Typically, the dependency is packaged with your compiled code but is not used during compilation.

➤ **testCompile**—Declares dependencies that are only required during the compilation of test source but will be left out while running the app.

➤ **testRuntime**—Declares dependencies that would be required while running the tests. Once again, they will be left out while running the app.

External Dependencies

Working with external dependencies might be the most important offering of build systems. Unlike local dependencies, external dependencies are available on repositories.

The most common approaches for dealing with external dependencies are as follows:

➤ Committing compiled binaries, which result in waste of disk space in source control, waste of network resources during commits, and waste of both when upgrading to a newer version.

➤ Copying/cloning source code into project, which results in a copy/paste fork of the target library. This approach will result in a project that is very hard to upgrade and is prone to cloning the bugs in the project.

Gradle resolves external dependencies within given repositories, either public or private. Gradle allows you to work with a range of versions of the dependency, or you can target the specific version you want to work with. In addition to this flexibility, Gradle also offers much simpler syntax to declare dependencies when compared to XML-based Maven syntax.

A typical Gradle dependency is declared with the library name followed by the version number. The following code snippet adds a supported library as a dependency:

```
dependencies {
    compile "com.android.support:support-v4:+"
}
```

The "+" character in the example tells Gradle that any version of support library is okay for the project. In such a case, Gradle will look for the most recent available version of the given project.

However, most of the time you need to declare a specific version of the target library to ensure compatibility and reproducibility. For example, to have Gradle download version 23.1.0 of the target library, type the version number as shown in the following code:

```
dependencies {
    compile "com.android.support:support-v4:23.1.0"
}
```

On the other hand, you may be looking to get minor version updates while still using a major version. For example, you might want the most recent update based on version 23.1 (i.e., 23.1.X). Once again Gradle lets you to use the "+" character for fine-tuning version numbers.

```
dependencies {
    compile "com.android.support:support-v4:23.1.+"
}
```

This example will retrieve the most recent support library based on version 23.1 but will not move to version 24 even if it is available. You can use the "+" sign for any digit or digits in the version number.

Although using Gradle is very easy and straightforward, you may need to have more control over the transitive dependencies. Gradle dependencies introduce their own dependencies, which would either form a tree or graph until a dependency does not need another dependency.

Usually transitive dependencies, which form a tree structure, do not impose any problem because each dependency has only one parent dependency. However if the transitive dependencies form a graph in which one dependency has more than one parent that requires that dependency, you may need to tweak dependency settings in order to provide the most suitable version for the needed dependency. Let's assume your project has two dependencies, A and B, which both require the dependency of C. If either A or B declares an incompatible version of C for the other, you would need to exclude the dependency from the graph.

This may also be an issue if your project already has a newer version of a dependency that is needed by another dependency. In the following code example, the project uses support library v4 23.1; let's assume `dependencyA` introduces an older version of the given support library.

```
dependencies {
    compile "com.android.support:support-v4:23.1.+"
    compile ("com.dependencyA:1.+") {
        exclude group: 'com.android.support', module: ' support-v4'
    }
}
```

This way, you ask `dependencyA` not to include `support-v4` because you know a newer version is already there.

Local Dependencies

As a best practice, you would need to upload jar or aar dependencies to private repositories even if they are not available on public repositories. However, if you still need to add a local jar or aar file as a dependency, you can point to the local library within parenthesis.

```
dependencies {
    compile "com.android.support:support-v4:23.1.+"
    compile files ("com.dependencyA_local.jar")
}
```

> **NOTE** *Having local binary file library dependencies defeats the point of dependency management and build systems and makes Gradle an unnecessary level of complexity over your project. With hardcoded local binary files, your project would never be easy to run in a new environment and would require custom configuration and setup.*

Although having a local binary file dependency is highly discouraged, you may need local modules, which already exist in source control, as dependencies. Gradle can easily declare dependencies between modules. The following example declares `moduleA` as a dependency of the project.

```
dependencies {
    compile "com.android.support:support-v4:23.1.+"
    compile :com.moduleA
}
```

Real projects might have a mixture of local module dependencies and dependencies from repositories.

Legacy Maven Dependencies

On occasion, you may not be able to find a Gradle reference to a dependency, but you can find the Maven reference. This used to be a common problem in the early days of the Gradle–Android flirtation.

Converting a Maven reference into a Gradle reference is pretty easy and straightforward. The example in Listing 6-3 declares for `log4j-api` and `log4j-core` version 2.4.1.

LISTING 6-3: Setting dependency version

```
<dependencies>
    <dependency>
        <groupId>org.apache.logging.log4j</groupId>
```

```
            <artifactId>log4j-api</artifactId>
        <version>2.4.1</version>
    </dependency>
    <dependency>
        <groupId>org.apache.logging.log4j</groupId>
        <artifactId>log4j-core</artifactId>
        <version>2.4.1</version>
    </dependency>
</dependencies>
```

To convert a Maven dependency to Gradle, you must start with mapping the `groupId` with a group, followed by mapping name with name and, finally, version with version, as shown in the following code.

```
dependencies {
    compile group: 'org.apache.logging.log4j', name: 'log4j-api', version: '2.4.1'
    compile group: 'org.apache.logging.log4j', name: 'log4j-core', version: '2.4.1'
}
```

Gradle also offers a simpler syntax, which allows a colon (:) to be used between each property without using the property names.

```
dependencies {
    compile 'org.apache.logging.log4j:log4j-api:2.4.1'
    compile 'org.apache.logging.log4j:log4j-core:2.4.1'
}
```

The ":" notation is the most accepted and widely used dependency declaration syntax in the Android ecosystem. Once you get used to Gradle dependency syntax, you could easily convert any Maven-based project or dependency to Gradle.

ANDROID PLUGIN FOR GRADLE

Gradle is great but what makes it better for Android is the Android Plugin for Gradle. So far in this chapter, you have used many features and properties of Android Plugin for Gradle. This section covers Android Plugin for Gradle in detail.

The new Android build system comes with Android Plugin for Gradle integrated with Android Studio. It can also be run independently, so it can easily be integrated with continuous integration servers. Either way, the build system will build the same APK described in the build.gradle file.

Configuring Android Plugin for Gradle

The "Anatomy of Gradle" section earlier in this chapter covered the basics of the Android Plugin for Gradle. As an Android developer, you may never develop and build applications without customizing the Android plugin, although it introduces many great capabilities without customization.

Build Configuration

The build.gradle file holds the build configuration for your project. You have already seen how to add dependencies, but the Android plugin for Gradle offers much beyond that.

The Android Plugin can control and configure the following items in your project:

➤ **Dependencies**—Dependency management is an important part of build systems, which enable dynamic, versioned, and transitive dependency management. We covered dependency management earlier in this chapter.

➤ **Android Manifest options**—Android Manifest is the heart of every Android application. The most trivial configuration details, such as application ID, supported compile/target/minimum SDK version, and application version info, will end up in `AndroidManifest.xml`.

➤ **Build type**—The Android build system is designed to build different binaries depending on your platform or application properties. This gives you the flexibility to build different applications or versions from one code base as well as different build options of the same code, such as debuggable or obfuscated builds.

➤ **Signing**—Applications need to be signed to be eligible to upload to the Google Play Store. By configuring signing settings, the build system can build ready-to-publish APKs without further user interaction.

➤ **Testing**—The build system can run your test during build and also package an APK file containing the test resources in your project.

➤ **ProGuard**—The flexibility of running in a virtual machine such as JVM, ART, Dalvik, and so on introduces easy-to-obfuscate portable byte code instructions. Prior to ProGuard, many Android applications suffered from decompilation and reverse engineering. ProGuard obfuscation not only is necessary for security but also shrinks the final APK size because variable, method, and class names are shortened.

Build Tasks

The Android build system is based on a set of hierarchical build tasks, which invoke child tasks in order to complete the whole build flow.

The following items are the top-level build tasks described by the Android build system.

➤ **Assemble**—Builds the project output, including code generation tasks and compilation

➤ **Check**—Runs checks (such as lint) and tests

➤ **Build**—Runs both assemble and check

➤ **Clean**—Cleans up the project

Flavors

Flavors, or build variants, are a flexible option provided by the Android build system. By default, each app comes with two different flavors: debug and release.

The following additional flavors can be defined for different purposes:

➤ Variations of an app such as free, demo, or paid versions.

➤ Apps with different app IDs from the same code base. Gradle is flexible enough to describe flavor-specific source and resource folders.

➤ Binaries for different CPU architectures such as ARM, x86, or MIPS.

To create a new flavor, you need to add your flavor definitions to the build.gradle file. Listing 6-4 declares two flavor versions for your app: demo (a free version) and full (the paid version).

LISTING 6-4: Gradle script with flavors

```
productFlavors {
    demo {
        applicationId = "com.expertandroid.chapter6.demo"
    }


    full {
        applicationId = " com.expertandroid.chapter6.full"
    }
}
```

You can also add a flavor by selecting the Edit flavors option from the Build menu. Click the plus (+) sign to define a new flavor. The new flavor with a default name will be created with empty options such application id, min sdk, and so on, which can be used to override the settings of the application defaults.

You have created a flavor that can be used while packaging your app and because you changed the app ID of both apps, they can be deployed to the Play Store as different apps. Now let's look at the changes needed for different app IDs.

First let's add a source file for each flavor:

1. Navigate to the src folder under your application and create two folders, demo and full, in addition to the main folder that's already there.

2. Create a folder named java and place your default package structure inside that. Each folder named for a flavor will inherit all the code inside the main folder but will also add the code inside its own folder.

3. Create a class with a constant field `demo` in the demo folder and `full` in the full folder.

4. Now it is time to select a flavor and use our flavor-specific code. Click the Select build variant option from the Build menu to open the window shown in Figure 6-1. Select demo from the bottom left of your IDE.

FIGURE 6-1: Build variant selection

Notice that the demo folder turned blue, indicating that you can use the demo flavor code in the `src/main` folder. If you select full, you can use the code inside the `src/full` folder.

5. Add different resources in your flavors.

Create a `rex/drawable` folder inside each flavor and copy a different `ic_launcher.png` to each to override the default icon. Notice the small yellow icon on the res folder, which shows that it's part of the active app's resources.

With the help of flavors, you can customize anything between builds, such as app ID, sources, resources, SDK version, UI layouts, assets—basically anything inside the main and flavor folders.

ProGuard

ProGuard is another great feature integrated into the Android build system. ProGuard is a tool for both security and performance. Before ProGuard, most Android applications were unprotected against decompilation and reverse engineering. ProGuard obfuscates your code by renaming classes, methods, and fields and removing unused code. The resulting APK is not only harder to reverse engineer but also smaller in size.

> **NOTE** *Although ProGuard is very easy to enable, it doesn't always come for free. If you use external libraries, you must check their ProGuard configuration in order to not break functionality. Usually, there's either no configuration or just a few lines to keep several class and method names untouched.*

ProGuard is enabled by default but only for the release version of your app. To enable ProGuard, the `minifyEnabled` property must be set to `true` in `buildTypes`.

```
buildTypes {
    release {
        minifyEnabled true
        proguardFiles getDefaultProguardFile('proguard-android.txt'),
        'proguard-rules.pro'
    }
}
```

> **WARNING** *You may need to configure ProGuard to keep specific method or class names, especially if your execution flow relies on reflection. Reflection works by looking up a method by its name, as represented in a string. Because the obfuscated name of a method by ProGuard is impossible to guess, ProGuard will most likely break reflection code. The same also applies for a method called from JNI.*

To configure ProGuard, you have two options. Android Studio adds `proguard-rules.txt` to the root of the project at project creation. This configuration file holds global ProGuard settings for whole modules in your project. For module-based configuration, `proguard-rules.pro` can be used.

To configure ProGuard not to obfuscate some part of the project, use the `-keep` option. Any class, interface, method, or field can be kept out of obfuscation with the `-keep` option.

```
-keep class com.expertandroid.chapter6.MyActivity
```

Listing 6-5 shows ProGuard options for the popular okhttp library from Square. To keep attributes, use `-keepattributes Signature` and `-keepatributes *Annotation*`. Refer to the ProGuard documentation for the full list of attribute settings. To keep classes and interfaces out of obfuscation, use `keep class` and `keep interface`.

LISTING 6-5: ProGuard options

```
-keepattributes Signature
-keepattributes *Annotation*
-keep class com.squareup.okhttp.** { *; }
-keep interface com.squareup.okhttp.** { *; }
-dontwarn com.squareup.okhttp.**
```

The previous section in this chapter covered product flavors. ProGuard also supports flavor-specific configuration. To add a new ProGuard configuration to the previous flavor example, you can add `full-rules.pro`.

```
productFlavors {
    demo {
        applicationId = "com.experandroid.chapter6.demo"
    }

    full {
        applicationId = " com.experandroid.chapter6.full"
        proguardFile 'full-rules.pro'
    }
}
```

Using ProGuard is essential for applications that will be released to the public. ProGuard not only protects your code against reverse engineering but also helps with securing your app.

Automated Tests

As mentioned previously in this chapter, Gradle executes all tests during the check task. The Android build system will execute both Android tests (built on JUnit) and JUnit tests. Chapter 8 covers testing and integrating tests; Chapter 10 covers Gradle and continuous integration.

GRADLE PLUGINS

The Android Plugin for Gradle is basically a Gradle plugin. Gradle plugins can be written with Java, Scala, and, of course, with Groovy. Each plugin can be put to work using the `apply` keyword with your plugin's name.

Writing Your Own Gradle Plugin

Writing a plugin of your own can customize the build process the way you want and is surprisingly something very easy to achieve.

Gradle plugins implement the `Plugin<Project>` interface; the `apply (Project p)` method needs to be implemented. As you might guess from the syntax, the targeted project is passed as a parameter to the `apply` method. Listing 6-6 adds the task `customTask` to your project, which currently only prints a log about starting the execution.

LISTING 6-6: Adding a custom task to a Gradle plugin

```
class CustomPlugin implement Plugin<Project> {

    void apply(Project p) {
        project.task('customTask') << {
            Log.info("Starting custom task...")
        }
    }
}
```

Now that your plugin is ready, you need to call `apply` to use it.

```
Apply plugin: CustomPlugin
```

Once the build process has executed, your plugin will run and print your log message. Alternatively, you can choose to run Gradle from the command line.

```
Gradle -q customTask
```

Let's add some more functionality to your plugin. Previously, you implemented different build flavors. Listing 6-7 will list all product flavors declared in your project.

LISTING 6-7: Listing all product flavors

```
class CustomPlugin implement Plugin<Project> {

    void apply(Project p) {
        project.task('customTask') << {
            Log.info("Starting custom task...")

            //check if this is an android project
            if (AndroidPluginTools.hasAndroidPlugin(p)){
                def flavors = p.android.productFlavors*.name
                for (String f in flavor){
                    Log.info("Product Flavor $f is found")
                }
            }
        }
    }
}
```

At this point you've added your plugin source code to the build script. This a very simple way to add a new plugin, but your new plugin is only available in your project. To promote reusability, create a separate project for the plugin that will be packaged as a jar and can easily be added to other projects.

Extending Android Plugin for Gradle

Extending the Android plugin can be useful and painful at the same time. The Android Plugin is just another Gradle plugin and is subject to change. Be aware that any change to the Android Plugin for Gradle may break your plugin's functionality. Listing 6-8 simply extends the Android plugin while displaying a simple log message.

LISTING 6-8: Display log message for a Gradle plugin

```
class CustomPlugin implement Plugin<Project> {

    void apply(Project p) {
        project.plugins.apply(AndroidPlugin.class)
        project.task('extendedAndroidPlugin') << {

            Log.info("Android Plugin is about to start")
        }
    }
}
```

Another great way to extend the Android plugin is to use `afterEvaluate`, which adds the defined closures to the end of the configuration phase. For example, let's say you want to create a report after running your extended task example:

```
afterEvaluate { project ->
  project.tasks.extendedAndroidPlugin << {
      println 'Your lint report is being generated'
  }
}
```

`afterEvaluate` can be used for adding hooks into any tasks. The execution order is based on first-in first-out, and the plugin does not have any control on the execution order.

SUMMARY

This chapter dug into some of the specifics of Gradle. We started with the basic syntax of Gradle and then focused on how to manage remote, local, and even Maven dependencies through Gradle.

In our exploration of the Android Plugin for Gradle, we showed you how to change its configuration, control the build tasks, and create flavors for different build settings from the same code base. Next, we moved to another important topic, ProGuard, and showed how to configure ProGuard for specific needs.

Finally, we covered the Gradle plugin system by showing how to write a Gradle plugin as well as extending Android Plugin for Gradle.

Entire books have been written about Gradle; this chapter's coverage really just scratches the surface of what is possible with advanced knowledge of the Groovy and Gradle lifecycles

7

Multi-Module Projects

WHAT'S IN THIS CHAPTER?

➤ Adding modules

➤ Phone & Tablet module

➤ Android Library module

➤ Android Wear module

➤ Android TV module

➤ Glass module

➤ Android Auto module

➤ Google Cloud module

➤ Importing modules

➤ Removing modules

Previous chapters dealt with general concepts of application development in Android Studio; this chapter covers the capabilities of Android Studio to work on multiple modules in your Android project.

In addition to the core module you created, you will eventually need additional modules when you want to add support for other Android devices. For example, you might want to add Wear, TV, or Auto modules in your Android Studio project, or even third-party Android libraries.

This chapter explores the details of adding Phone & Tablet modules, Android and Java libraries, and Wear, TV, Glass, and Google Cloud modules. Then the chapter covers importing Gradle and Eclipse projects. AAR and JAR packages are covered as well to help you understand how to include them in your Android Studio projects.

Android Studio works on a module-based project structure, which means that it can handle multiple modules in one project. Having multiple modules in an Android Studio project enables you to work on one project instead of multiple projects so your development team can create a more organized application.

Prior to Android Studio, Eclipse handled multi-module projects with a workspace concept. Eclipse projects do not need to reside in the same project folder to be included in the same project setup. However, this approach relied on Eclipse project dependencies for the build process.

A better organized approach uses Maven. Although Maven is a *de facto* standard in build and dependency management in Java projects, it was not always supported in Android projects. The introduction of Gradle as the default build system for Android solved this huge problem in standardization of dependency management.

ADDING MODULES TO ANDROID PROJECT

In Chapter 3, you learned how to create a new project with multiple modules. In this section, you learn how to add new modules to an existing Android Project.

We created a new project named ChapterSeven to work on modules. You can recreate that project, create a new project, or work on a project you previously created.

When your project is ready, find the actions from the Android Studio menu to add a new module. Either right-click the project pane or open the File menu and select New ⇨ Module to start adding a new module to your project.

After clicking the Module option, the wizard shown in Figure 7-1 opens. Select the type of module to add to your project.

By default, you will see ten module actions in the window; you can select to either add a new module or import an existing module to your project.

After selecting the module, Gradle files will be auto-generated to handle a multiple module build and release process. The following sections cover the auto-generated files, folders, new Java packages, and Gradle, manifest, and resource files in more detail for each module type.

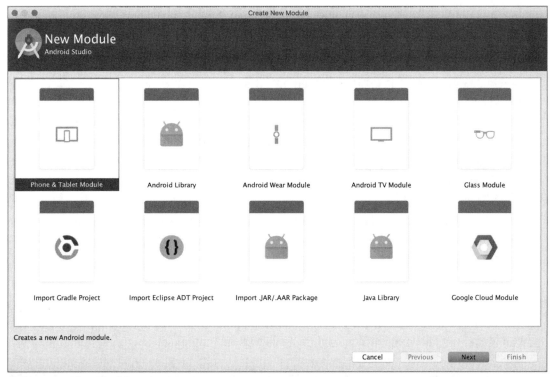

FIGURE 7-1: New Module selection

PHONE & TABLET MODULE

Most of the time, the Phone & Tablet module (also called the Android module) is the core module in a project. While it might not be common to add multiple Android modules together, in certain cases you may require a second or third Android module in addition to your core module to generate multiple APKs with different properties.

For example, you can develop an Android application that supports multiple screen sizes with configuration of dimensions and a good design of XML resources, bitmap resources with multiple resolutions, and support libraries. However, in that case, the APK files get a lot larger and create significant cost for the user to download and install. To avoid that, you can create multiple modules to generate individual APKs for specified devices.

Alternatively, when you develop an application with trial, demo, enterprise, consumer, or paid versions you would need modules with different APKs to distribute in the Google Play Store: one module with limited access and another with unlimited access to all resources and activities you have developed.

You can also develop a tutorial application, managing the new lightweight application within the project and generating binaries and APKs with a single build system.

There are other situations where you may want to manage multiple Phone & Tablet modules in the same project, such as when starting the project with the Wear module and adding a new Phone & Tablet module and so on.

Managing multiple modules in Android Studio gives you the opportunity to configure Gradle for handling dependencies between modules, use a shared module's resources or libraries, and generate multiple APKs at the same time in the same project directory. Using Gradle eases the development process for multiple teams working on a complex application, and helps the transition to continuous integration.

If you select the Phone & Tablet module (refer to Figure 7-1), you need to configure module name and activity type as you would when creating a new Phone & Tablet project. Figure 7-2 shows that we added a new Phone & Tablet module named chapterseventrial.

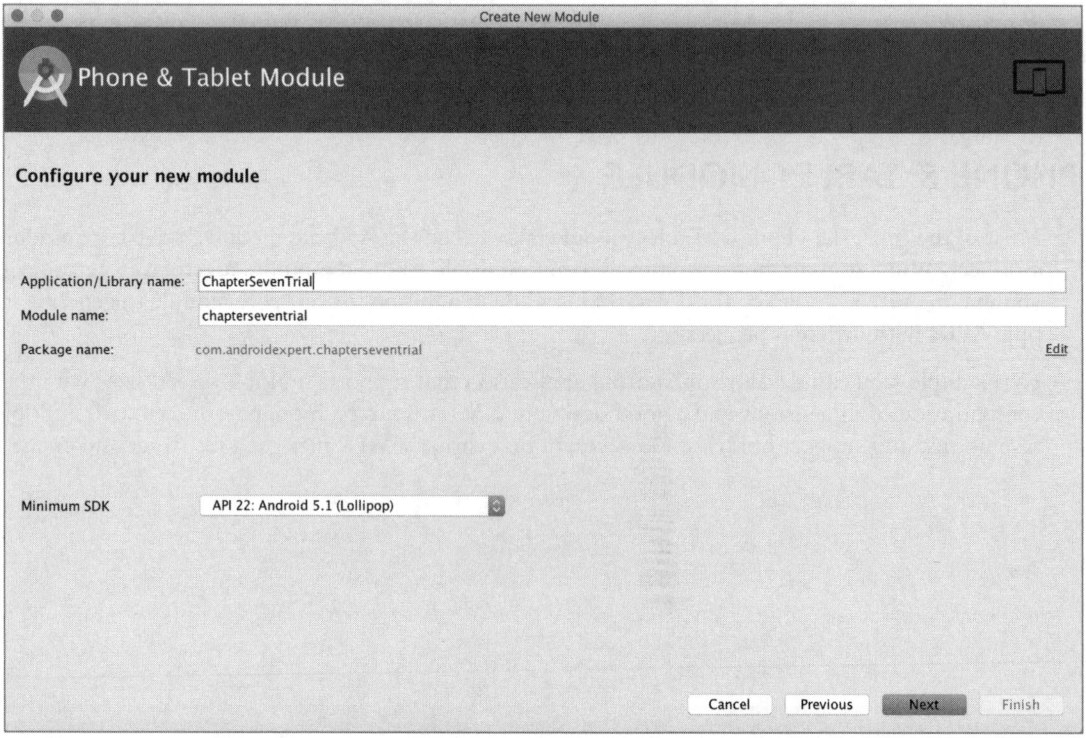

FIGURE 7-2: Configuring a new Phone & Tablet module

After naming the module, you follow steps that are similar to those for creating a new Android project. After naming and configuration, the module's files and folders are generated and the chapterseventrial module appears in the Android Project view as a separate module, as shown in Figure 7-3.

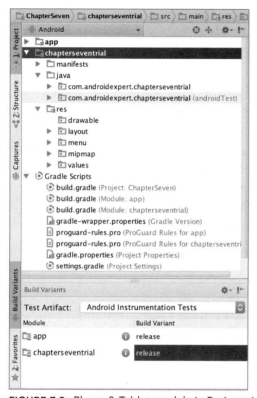

FIGURE 7-3: Phone & Tablet module in Project view

As shown in Figure 7-3, the new module has its own folder, resources, Java source code, and manifest and Gradle files for configurations. The Build Variants window at the bottom shows the build type of the modules to configure and generate multiple APKs.

Build configurations can be made in Gradle or from the Project Structure window (Command+semicolon [;] on Mac, Ctrl+Alt+Shift+S on Windows). From the Project Structure window, you can configure the module for versions, APK signing properties, build types, and flavors and their dependency to each other.

You can initiate a project build by selecting Make Project from the Build menu, or you can just build the module by selecting Make Module *<modulename>*.

When you select Generate APKs from the Build menu, APKs will be generated under module's `build/outputs/apk` folder. You can navigate to the directory from the folders where you store the Android Studio project such as `/Users/username/AndroidStudioProjects/ChapterSeven/chapterseventrial/build/outputs/apk` on Mac OS X or `C:\Users\username\Projects\AndroidStudioProjects\ChapterSeven\chapterseventrial\build\outputs\apk` on Windows.

> **TIP** *If you have multiple Android modules for similar applications, you should configure your Android manifest so that the APKs are identified correctly when published to Google Play.*

ANDROID LIBRARIES

The Android Library module contains the shareable Android source code and resources that can be referenced by other modules in an Android project.

The Android Library module is useful when you want to reuse and share a code base or XML resources within your project modules. This can be either a private library developed by yourself, or a third-party library imported into your project.

To add and develop your own Android Library module, select Android Library from the list shown in Figure 7-1. When you proceed to configuration, the module wizard asks only the library's name.

When you finish adding the library, you can see its package, folders, and Gradle file under the project folder, as shown in Figure 7-4.

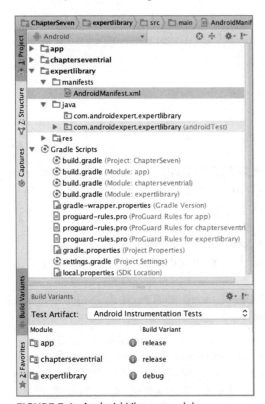

FIGURE 7-4: Android Library module

Android libraries are useful when you are developing multiple application projects or multiple APKs and want to use a shared resource to lower implementation time and preserve continuity between modules and projects.

Working with Android Libraries

An Android library is not a runnable application on its own; it doesn't generate an installable APK but generates a shareable AAR package. An Android library acts as a shareable resource that is loaded and fetched by the apps using the library.

Adding an Android library is not enough to enable you to start using it in other modules. Before using an Android library in your application module, you should add it from Gradle or from the module settings window as a dependency to the corresponding module.

Dependencies can be configured through an interface from the Project Structure window, which you can open from the File menu. Alternatively, you can access the Project Structure window by right-clicking the module in Project view and selecting Open Module Settings.

Select the module that requires an Android library, and then navigate to the Dependencies tab, as shown in Figure 7-5.

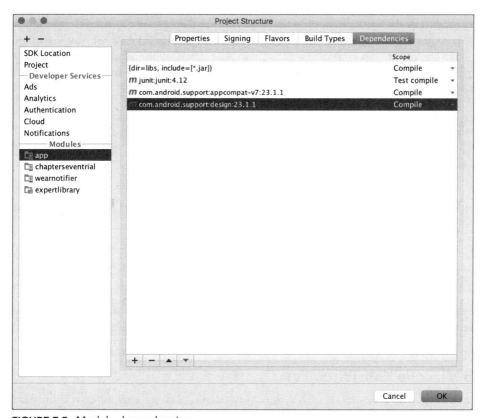

FIGURE 7-5: Module dependencies

Next, click the plus button at the bottom to add the module dependency (see Figure 7-6). When you finish adding the module dependency, the expertlibrary module will appear in the list of dependencies.

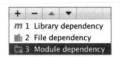

FIGURE 7-6: Add a dependency

You can also add a dependency directly from a module's build.gradle file. To do that, add the `compile project(':androidlibrary')` line to the dependency area and sync the project to make your module load the library.

We added the expertlibrary module together with auto-generated dependencies to our project's Gradle file dependencies section, as shown in Listing 7-1.

LISTING 7-1: Dependencies build.gradle app module

```
dependencies {
    compile fileTree(dir: 'libs', include: ['*.jar'])
    testCompile 'junit:junit:4.12'
    compile 'com.android.support:appcompat-v7:23.1.1'
    compile 'com.android.support:design:23.1.1'
    compile project(':expertlibrary')
}
```

Now you are ready to use shared resources and Java classes in your module. When you start typing the class name in the editor, autocomplete will list matching item(s). Similarly, when you assign a resource to your layout view, you will see the shared resource in the list of resources when typing (or adding from the list).

> **NOTE** *If the dependency isn't defined, the Android library classes or resources are not accessible from the other modules. Gradle sync links the library to the module.*

Using Android Library Java Packages

To start using Java classes from an Android library, add a new Java file to the library package under the src folder. You can add as many Java classes as you need to create a shareable library and speed the development process.

When a dependency is set and a Java class is added to your Android Library module, you can use it in the corresponding module. When you are in the app module, type the name of the class you want to add and it will be ready for use.

Using Android Library XML Resources

Sharing resources between modules is really practical when you need to use your own defined color values in both your phone and wear apps. Using resources makes the development of user interface elements easier because you don't have to rewrite the same definition repeatedly for each module.

When you're finished with dependency configurations, add a new XML resource to the library. We added `colors.xml`, which was not present when the project was created. This XML file defines the colors' names and RGB values to use in all other modules.

Finally, go to your module's layout file to assign the color resource you want to use. If Gradle sync worked, you will see the resource in the list or in the XML editor. It will be listed right after you start typing.

Generating an AAR Package from an Android Library Module

As mentioned earlier in this section, an Android library is not an application but is loaded into the dependent modules from the provided binary dex file and is stored in the resources folders.

You may want to share your library with other people or make it open source and distribute it. The easier way to distribute your Android library is to share the output AAR file. The AAR file is similar to an APK file, which is a compressed file that includes the necessary content of an Android Library module for easy export and import. Having a standard file extension such as `.aar` helps developers recognize that the library is for Android applications, not for Java.

When you select the Build APK option from the Build menu, an AAR file is generated and placed in the module's `build/outputs/aar` directory. When you build our Android Library module, you will see the `expertlibrary-release.aar` or `expertlibrary-debug.aar` file in the output directory, as determined by the Build Variant configuration.

JAVA LIBRARIES

The Java Library module includes Java packages to link with your modules and Java classes to reuse as needed in your projects. Depending on the specific requirements of your project, an Android Library module can also be used for this, but as stated in the following note, that might not be the best solution.

> **NOTE** *The principle of separation of concerns means that it is better to cover core Java-related work in a Java library and Android API–related work in an Android library.*

Add a new Java library to your project by selecting New Module and then select Java Library from the list of modules (refer to Figure 7-1). The Create New Module wizard shown in Figure 7-7 will open.

For this configuration, name the library, which becomes the name of your library and module. Provide a name for the initial Java class you want to create in the library. You can edit the name of the package by clicking on the Edit link to the right of the Java package name line. The default name for the package is `com.example`, as shown in Figure 7-7.

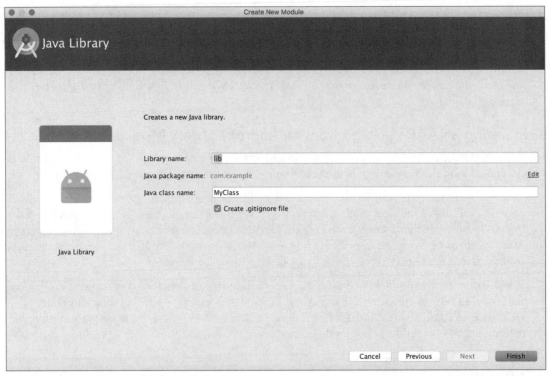

FIGURE 7-7: Java Library module configuration

As shown in Figure 7-8, the new Java Library module is shown in Project view after you click Finish and when Gradle sync completes.

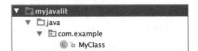

FIGURE 7-8: Java Library module in Project view

Like an Android library, a Java library must be added as a module dependency for the corresponding module. In order to reference a library in the module, you can edit the module's build.gradle file, as shown in Listing 7-1. You can also add the library from the Project Structure window by selecting the module and selecting the Dependencies tab as we did in the "Android Libraries" section (refer to Figure 7-6).

When you are done with dependency configuration, you are ready to use the Java classes in the module that the Java Library referenced. Java class names are autocompleted when you start typing them in the referenced module's Java source file.

ANDROID WEAR MODULE

Wearable technologies are getting better day-by-day and Android is the frontier for wearables, with multiple devices on the market running Android Wear. As a result, there is a large installable base online for a wearable Android application developed with the Android Wear API.

> **NOTE** *Currently, Wear devices refer to smart watches or bracelets.*

> **NOTE** *Although Google Glass is a wearable device, the Android Wear API doesn't work for Google Glass. Google Glass uses a separate module with a different API version and packages.*

In order to develop a wearable project, you can either start a standalone Android Wear project with Android Studio or add a new module to your existing Android project.

Having a Wear module in your project makes sense when you want to distribute your application's extended features, such as designing a basic user interface with touch or voice recognition for a Wear application that interacts with your main Android phone or tablet application.

After you select the Android Wear module from the Create New Module window shown in Figure 7-1, the configuration window asks for the name of the module and the initial wear activity and its name.

When you finish adding the Android Wear module, it appears as a separate module in the application project window.

During UI development, keep in mind that Wear projects use totally new form factors. However, you can share the color, theme, font, and text resources you generated in your Android library project between your Android Phone & Tablet project and Wear project.

After you add the Wear module with an Empty Activity, you will see that three layout files have been created: two for multiple form factors on wearable devices (round and rectangle), and one for the main activity (see Figure 7-9). The round and rectangle layouts should be designed according to their shape.

As shown in Figure 7-9, the module has its own resources, activity class, and Android manifest file. Although further enhancements and development are similar to Android application development, you should consider during the design and developing that you will deploy this application to far less powerful and smaller devices than a smartphone.

Also consider that your application will be running on a very small screen. That means that multi touch is hard to accomplish, and there is space for only a little information using a small font that might be hard to read.

Google provides plenty of information about Wear. Visit the following URL for Android Wear application guidelines and training: `http://developer.android.com/wear/index.html`.

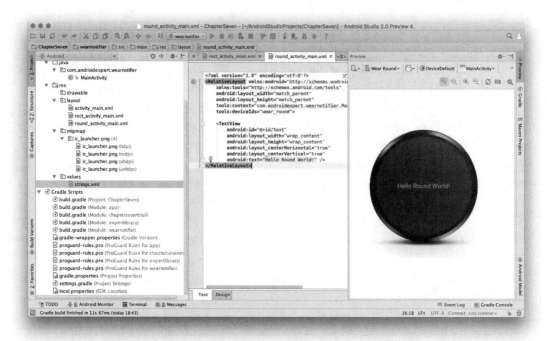

FIGURE 7-9: Wear module preview

Running and Debugging an Android Wear Module

Android Wear devices are not as common as smartphones, but they will be eventually, and they will be cheap and easy to afford. For now, you can take advantage of virtual devices to run and debug Android Wear modules.

Creating an Android Wear virtual device is similar to creating a phone or tablet device. (See Chapter 2 if you need a refresher on creating a new virtual device.) Instead of selecting Phone, as in Chapter 2, here you select Wear from the virtual device category as shown in Figure 7-10.

After selecting the hardware profile, select the Android version to run on the Android Wear virtual device. We selected Android 6.0, Marshmallow, for this example.

Finally, select the device configuration to finish creating a Wear virtual device, as shown in Figure 7-11.

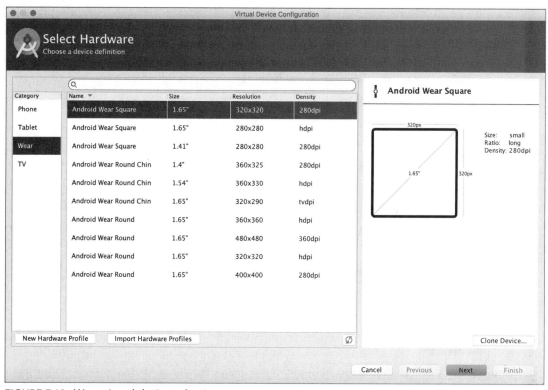

FIGURE 7-10: Wear virtual device selection

Now, you can run the selected module: Go to the Run menu and select the Run action instead of Run <*name of default module*>, or press Control+Option+R on Mac, Alt+Shift+F10 for Windows. This action will list the runnable modules, as shown in Figure 7-12.

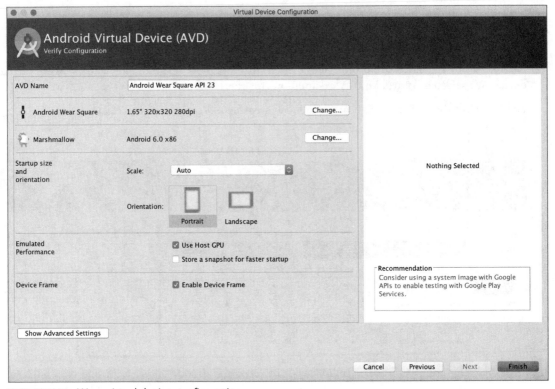

FIGURE 7-11: Wear virtual device configuration

FIGURE 7-12: List of runnable modules

> **NOTE** *If there isn't a Wear device connected to your development machine or an AVD created on your development environment, your Wear module will not launch. Even if you select the Phone emulator, your Wear module will not be installed on the device.*

Launching the Wear module on a square Wear device will install the generated APK of the Wear module and show the Wear app, as in Figure 7-13.

FIGURE 7-13: Wear app in Android Wear AVD

To debug your app, select the Debug action or press Control+Option+D on Mac, Alt+Shift+F9 for Windows to show the debug windows.

Building APKs with Android Wear Support

After adding a Wear module, you need to make sure the build process handles the APK generation correctly.

Chapter 4 covered the steps to generate a signed APK. Follow those steps to assign a signature to a Wear module. Android Studio handles all modules separately. Select the Build APK option or the Generate Signed APK option from the Build menu to generate the APKs. Android Studio will open a notification window when it is finished, as shown in Figure 7-14. Follow the link at the bottom to open path in the file browser.

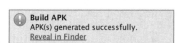

FIGURE 7-14: APK generation completed notification

When the build process is finished and you navigate to the project folder, you will see the APK under the module's `build/outputs/apk` directory.

ANDROID TV MODULE

Although Android TV might seem new in the market, it has a long history. It was first announced as Google TV at Google I/O 2010. The first available devices were from Logitech and Sony. The first generation of Google TV devices were all designed on Intel's x86 platform, although the second generation of Google TV ran on ARM devices.

Another effort to bring Android to a big screen was Nexus Q, announced at Google I/O 2012. Nexus Q was a high-quality device with an integrated amplifier, NFC support, and nice design.

However, the price tag was three times more than similar devices like Apple TV so it never really got far from being a prototype device.

Two years later, at Google I/O 2014, a new Android-based device was announced—Android TV. The first model, labeled ADT-1, was a developer-only device. A few months later, an Intel-based Nexus Player was announced, followed by several TV manufacturers offering Android TV capabilities integrated in their TV sets.

Android TV is creating its market presence, and developers are becoming more focused on the Android TV platform, extending their applications for TV use. These applications can be game or media players offering a "leanback" user experience.

> **NOTE** *Google has released lots of developer resources for Android TV projects because its design requires a different approach than small devices like phone or wear. The following URL provides information about Android TV application design guidelines and standards:* `https://www.android.com/tv/`.

Adding a new Android TV module is similar to adding a Wear module. You should first name the module and then select an activity. Android TV has only one template activity to select when you configure the module. It is not a simple template to manage at the beginning for a simple application but it is a stable template and can run on an Android TV device immediately.

The activity customization window asks for many additional configurations to add multiple fragments and activities, as shown in Figure 7-15.

When you click Finish without changing the activity and fragment names, the Project view will show many Java classes and resources generated under the `res/drawable` folder, as shown in Figure 7-16. We named our TV module MirrorApp, a common use case for TV applications to mirror the current content of a smartphone or desktop application to a TV screen.

The Android TV activity template is focused on media playback and adds Java classes and resources to manage video playback seamlessly in the application.

The Android TV manifest is a little different than other applications because of its input method. Because TVs don't have a touch input, the launcher `Intent` uses the `LEANBACK_LAUNCHER` category flag to be accessible from the Android TV launcher and signify to the Google Play Store that it's compatible with TVs.

The auto-generated Android TV module manifest file is shown in Listing 7-2. It disables the touch screen input requirement and requires the leanback feature for the application, making this module a TV-only application.

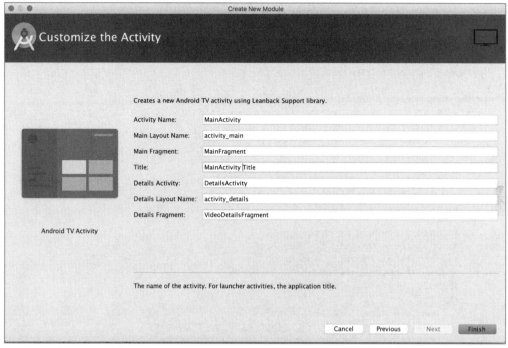

FIGURE 7-15: Android TV module activity configuration

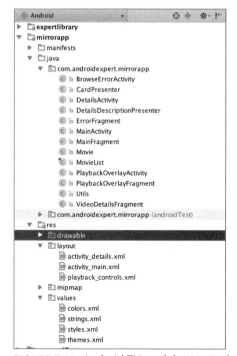

FIGURE 7-16: Android TV module view in the project window

LISTING 7-2: Android TV module manifest file content

```xml
<?xml version="1.0" encoding="utf-8"?>
<manifest xmlns:android="http://schemas.android.com/apk/res/android"
  package="com.androidexpert.mirrorapp">

  <uses-permission android:name="android.permission.INTERNET" />
  <uses-permission android:name="android.permission.RECORD_AUDIO" />
  <uses-feature
      android:name="android.hardware.touchscreen"
      android:required="false" />
  <uses-feature
      android:name="android.software.leanback"
      android:required="true" />

  <application
      android:allowBackup="true"
      android:icon="@mipmap/ic_launcher"
      android:label="@string/app_name"
      android:supportsRtl="true"
      android:theme="@style/Theme.Leanback">
      <activity
          android:name=".MainActivity"
          android:banner="@drawable/app_icon_your_company"
          android:icon="@drawable/app_icon_your_company"
          android:label="@string/app_name"
          android:logo="@drawable/app_icon_your_company"
          android:screenOrientation="landscape">
          <intent-filter>
              <action android:name="android.intent.action.MAIN" />

              <category android:name="android.intent.category.LEANBACK_LAUNCHER" />
          </intent-filter>
      </activity>
      <activity android:name=".DetailsActivity" />
      <activity android:name=".PlaybackOverlayActivity" />
      <activity android:name=".BrowseErrorActivity" />
  </application>
</manifest>
```

Running and Debugging Android TV Modules

If you already have an Android TV device such as the Nexus Player, you can use it to run and debug your applications; otherwise, you will need an Android TV emulator to run and debug your TV module.

Let's create a TV emulator by opening AVD Manager and clicking New Virtual Device to select TV from the left panel. Then select the resolution (1920 x 1080) and density (xhdpi) of the TV emulator, as shown in Figure 7-17.

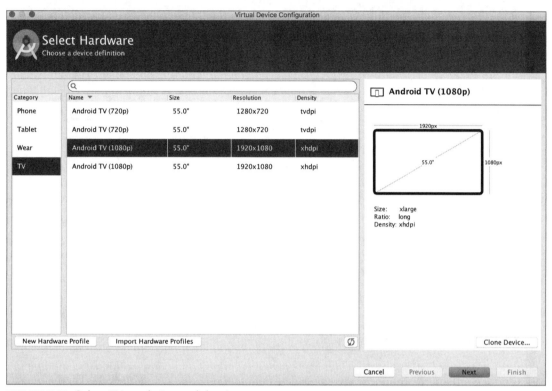

FIGURE 7-17: Select TV resolution and density

The next window lists existing Android TV images with the corresponding Android API version. Figure 7-18 shows that Android 6.0, Marshmallow, has been chosen as the TV emulator.

Right after you select the Android API version, click Next, review the virtual device on the next screen, click the Finish button, and you are done. Then you are ready to launch sample module `mir-rorapp` on the virtual Android TV.

To launch the TV module, follow the same steps as you have for other modules: Select Run from the Run menu or press Control+Option+R on a Mac, Alt+Shift+F10 on a Windows to select your TV module from the popup dialog box shown in Figure 7-19.

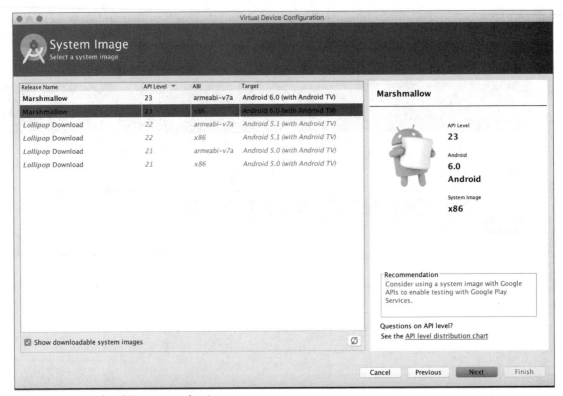

FIGURE 7-18: Android TV image selection

Next, select the recently created virtual device or your Nexus Player to launch the TV module. When the virtual device launches, the custom Android TV template application appears on the TV emulator, as shown in Figure 7-20. We didn't make any changes on the Android TV template activity so it will only show the auto-generated text and layout.

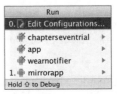

FIGURE 7-19: TV module selection

To display the Debug menu shown in Figure 7-21, click Debug from the Run menu, or press Control+Option+D on a Mac or Alt+Shift+F9 on Windows.

While developing a new TV application, you can follow the standards for Android application development, but the user interface is a lot different than the interface for smartphones and wear devices.

The user interface should be designed for a huge screen instead of small screen device. Users interact with applications with a remote control or a game pad. So, although core application development adheres to similar principles, the user experience design changes.

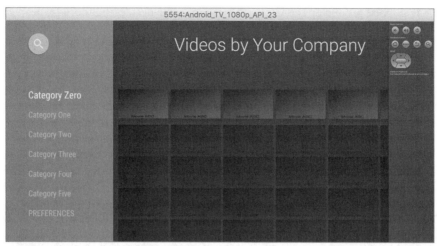

FIGURE 7-20: Running Android TV application on AVD

Building APKs for Android TV Modules

Android TV apps and modules are developed to run on Android TV devices so you need a separate APK file to publish them on Google Play and distribute your app to users.

In order to build an APK for the TV application, select Build APK from the Build menu to generate an unsigned APK.

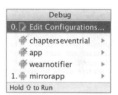

FIGURE 7-21: Android TV module debug selection

The Project Structure window can be used for configuring the TV module's Properties, Signing, Flavors, Build Types, and Dependencies.

If you've configured signing in the Project Structure window, you can populate the signed APK for the TV module by selecting the Generate Signed APK option from the Build menu. APK files are stored in the project's root folder, under the specified TV module's sub-folder, in `build/outputs/apk`.

GLASS MODULE

Google Glass is a wearable device first announced at Google I/O 2013. The first batch of devices was sold to a limited number of developers, called Glass Explorers. The Explorer program was later extended to more developers and continued until January 2015. The Glass Explorer program was discontinued, but Google announced that it is committed to newer versions of Glass.

Although Google has stopped producing the Google Glass prototype device, that doesn't mean that there will be no new devices available on the market. Also, you can still get access to Google Glass Preview API version 19 to create Glass modules based on the Glass Explorer edition.

> **NOTE** *You should download the Google Glass Preview API from SDK Manager. You can find the system image and API under Android 4.4.2 (API 19).*

After you select the Glass module shown in Figure 7-1, you will be prompted to name the Glass Module. As you can see in Figure 7-22, the Glass module for this project is named seefrommyglass.

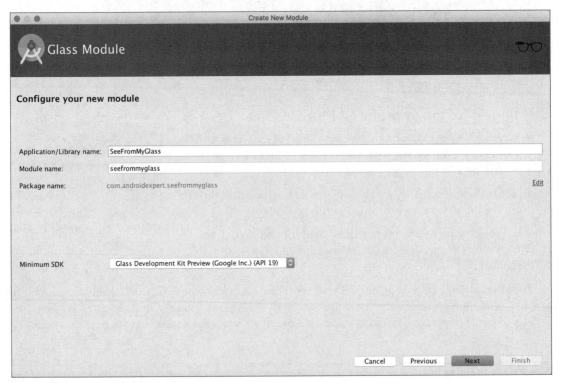

FIGURE 7-22: Glass module naming

After naming the Glass app, you will be prompted to select an activity type. The two selections to choose from are shown in Figure 7-23.

You can either pick Immersion Activity or Simple Live Card for the application. Immersions are Google Glass apps with a user interface. Live Cards are like widgets on Android phones, which appear together with a clock, like a notification on the screen.

Name the activity or card selection in the next window, and the Glass module is ready to build. All necessary files will be auto-generated by Android Studio right after the activity or live card is named.

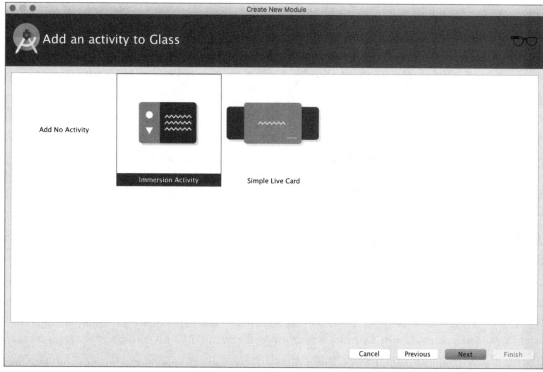

FIGURE 7-23: Glass module activity selection

Like all other module types, Glass is also an Android application so development of a Glass module is done following the standards and guidelines applicable to any Android applications, with some exceptions. The user experience design is different because user interaction with the Glass device is different than with other devices. Voice recognition is a dominant factor compared to touch to interact with applications on Glass.

> **REFERENCE** *Google provides design and development guidelines for Glass applications. The following resource can be used to learn more about Google Glass application development:* `https://developers.google.com/glass/design/principles`.

Running and Debugging a Glass Module

If you are lucky enough to have a Google Glass device, you can test your module on it. Select Run or Debug from the Run menu and select the Glass module from the dialog box that opens. Then select the hardware device on which to run the Glass module.

> **WARNING** *Unfortunately there is no Google Glass emulator to test your Google Glass application. You will need a Google Glass device or you can use the Google Mirror API Playground (which doesn't support the native API) for testing:* `https://developers.google.com/glass/tools-downloads/playground`.

Building APKs for Glass Module

Building APKs for a Glass module is not much different than it is for the other modules. When you click Build APK or Generate Signed APK from the Build menu, a Glass module APK will be generated in the Glass modules folder under the project directory in the `build/outputs/apk` directory.

ANDROID AUTO MODULE

Android Auto is the new kid on the block. There are no cars yet available with Android-equipped hardware. However, Google is very committed to Android Auto and is working with several car manufacturers. Google also demonstrated Android Auto simulators at Google I/O 2015.

Although Android Auto applications will be deployed on a separate platform, eventually Auto will be available as a module in Android Studio and will generate APKs to deploy applications on cars.

There isn't a defined module for Android Auto yet, but you can enable an Android module to work with Android Auto by adding a media or messaging Android service on your application after you create the Android project. Then you need to further configure your app so it works with Android Auto. You do this configuration in the module's Android manifest file, beginning with setting permissions, as in Listing 7-3.

LISTING 7-3: Android Manifest configuration permissions for Android Auto

```
<uses-permission android:name="com.google.android.c2dm.permission.RECEIVE" />
<permission
    android:name="com.example.gcm.permission.C2D_MESSAGE"
    android:protectionLevel="signature" />

<uses-permission android:name="com.example.gcm.permission.C2D_MESSAGE" />
```

In addition to Android Service extending `MediaBrowserService` and Android Manifest file permission configurations, you should add a new `automative_app_desc.xml` resource file for Auto enabling. This file, shown in Listing 7-4 and located in Android Auto's `res/xml` folder, allocates resources for your application. Then you need to configure the Android Manifest file so that your application uses the resource, as shown in Listing 7-5.

LISTING 7-4: automative_app_desc.xml

```xml
<?xml version="1.0" encoding="utf-8"?>
<automotiveApp>
    <uses name="media" />
</automotiveApp>
```

LISTING 7-5: Application metadata for Android Manifest

```xml
<meta-data
    android:name="com.google.android.gms.car.application"
    android:resource="@xml/automotive_app_desc" />
```

All the steps mentioned previously are auto-generated if you are creating a new project with Android Auto support. There will be two auto activities to select during project creation, as shown in Figure 7-24.

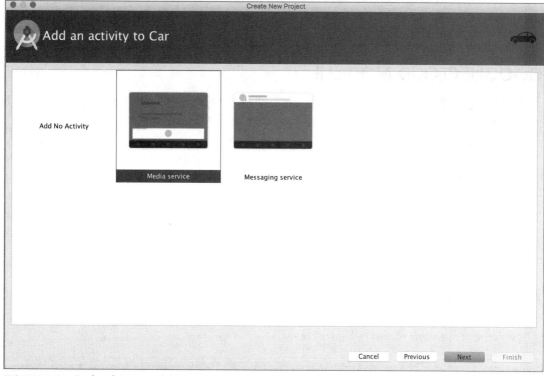

FIGURE 7-24: Android Auto activities

Figure 7-24 shows that Android Auto is enabled for:

➤ **Media Service**—Control media content through the Android Auto interface to play, stop, or skip media selections.

➤ **Messaging Service**—View and respond through the Android Auto interface.

You can add support for Android Auto but there isn't a virtual device yet for it. However, you can work on an Android Auto–enabled application using the Android Auto Desktop Head Unit Emulator. That can be installed with the Android SDK Manager, under the Extras section.

To test Android Auto features, you need to install the Android Auto app to your smartphone from the Google Play Store, which emulates an Auto device. Next, you need to configure ADB from a terminal to communicate with the Android Auto Desktop Head Unit Emulator's auto executable, which is found in the `AndroidSDKPath/sdk/extras/google` folder. There is no direct connection to Android Studio at this point, but it will certainly come in the future.

> **REFERENCE** *More Android Auto resources can be found at* `http://developer.android.com/training/auto/start/index.html`.

GOOGLE CLOUD MODULE

The Google Cloud module, shown previously in Figure 7-1, is not directly related to the Android API or devices but to the backend or cloud side of the Android application.

The Google Cloud module is used for developing and deploying backend applications to the Google App Engine and to scale your application so it can handle multiple users, store data, collect analytical information, and perform all other tasks one can do in the cloud.

Multi-module development capability is a powerful feature of Android Studio that provides more capability, and not only when dealing with Android and Java applications or libraries but also, with backend module development, extended features of Android applications with cloud connectivity. Having this capability helps developers manage the client- and server-side processes within Android Studio while also managing the build process and the deployment of properties.

Adding a New Google Cloud Module

Add the Google Cloud module by selecting it from the window shown in Figure 7-1. The configuration wizard opens, as shown in Figure 7-25.

> **NOTE** *To develop the Google Cloud module, make sure you've installed the Google APIs from the Android SDK Manager.*

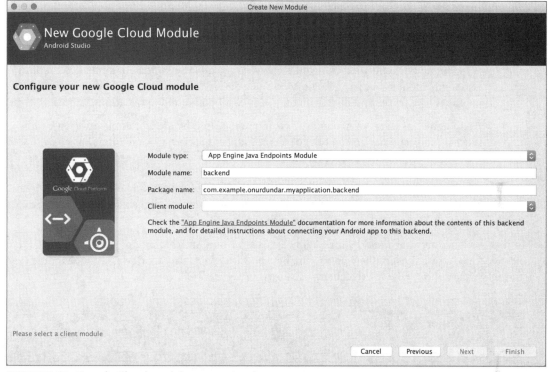

FIGURE 7-25: Google Cloud module setup wizard

The wizard displays four fields: Module type, Module name, Package name, and Client module. These fields are described in more detail in this section.

The list of module types consists of three types of Google Cloud module templates that can be added to a project, as shown in Figure 7-26.

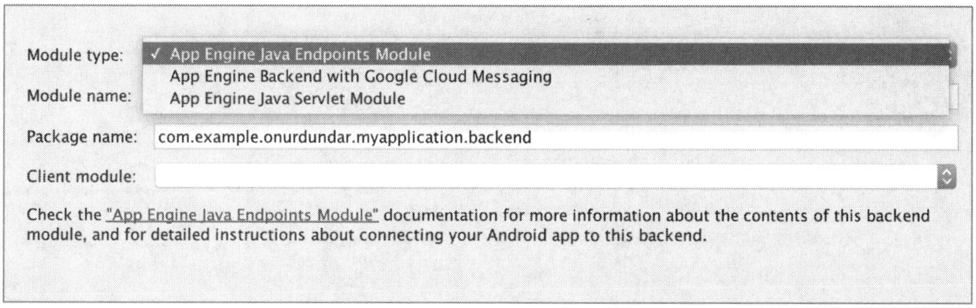

FIGURE 7-26: Google Cloud module types

These Google Cloud module types are:

➤ **App Engine Java Endpoints Module**—This is the module used to build the backend Restful API to handle requests and send data with the REST API.

➤ **App Engine Backend with Google Cloud Messaging**—This module is similar to the Java Endpoints module but has Google Messaging Service support to enable your app to send messages to a server and distribute messages to all connected applications.

➤ **App Engine Java Servlet Module**—This module is used by a client application to build requests and send data using `httpClient`. The Java Servlet module offers easier or simpler use cases than the Endpoints module.

Additional information can be accessed by clicking the link (blue text) under the selection boxes. The links take you to the module templates' Github home pages so you can see the detailed explanation of each module type and related source code.

In this section, you want to add a messaging service for our Android module, so name the Cloud module "messaging," select the App Engine Backend with Google Cloud Messaging Module type, and select the app module for the client, as shown in Figure 7-27.

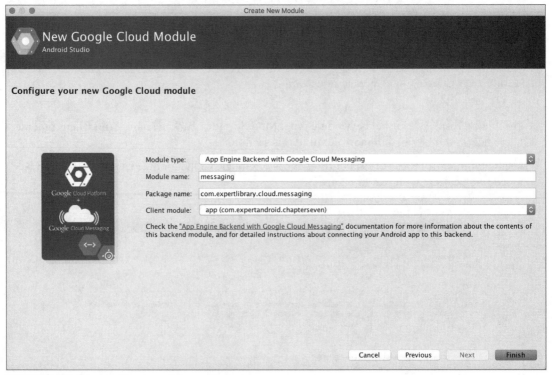

FIGURE 7-27: Google Cloud module configuration

After you click Finish and Gradle sync successfully finishes, you can see that the messaging module is added to Android Project view, as shown in Figure 7-28.

FIGURE 7-28: Google Cloud module after module creation

Running and Debugging a Google Cloud Module

Because the Google Cloud module has a different API, it needs to be handled by Android Studio separately. First build the project by selecting Make Project from the Build menu.

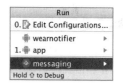

FIGURE 7-29: Runnable modules with Google Cloud module

If the build is successful, you can select Run or Debug from the Run menu to list runnable modules, as shown in Figure 7-29.

Selecting the messaging module will generate a runnable Cloud module and start a local App Engine Java Development server to run and debug the module locally. Before running, a message window shows the output of the process and enabled servers, as shown in Figure 7-30.

Figure 7-30 shows the output from the Google Cloud module (`http://localhost:8080`). There is a second URL that points to the admin page (`http://localhost:8080/_eh/admin`) for the server configuration and module monitoring. Let's navigate to that page to see that the module is running, as shown in Figure 7-31.

The links provided on the localhost page take you to additional information where you can learn more and further develop the module.

> **NOTE** *Further testing requires a Google Cloud Messaging API key to send and receive messages. You can follow up with the development and procedures at* `https://github.com/GoogleCloudPlatform/gradle-appengine-templates/tree/master/GcmEndpoints`*.*

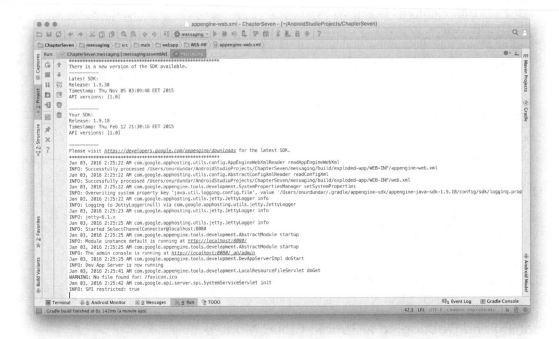

FIGURE 7-30: Message output for the Google Cloud module run

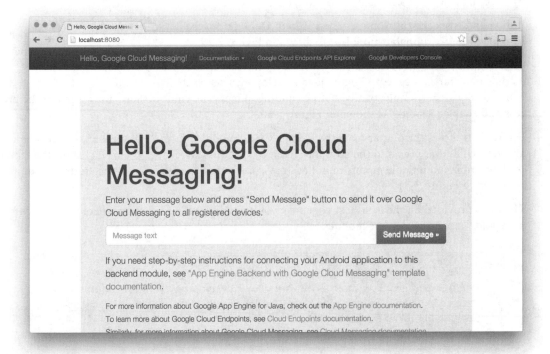

FIGURE 7-31: Google Cloud module launch from browser

IMPORTING MODULES

This section covers the inclusion of external modules, such as the Gradle project, the Eclipse ADT roject, and .JAR/.AAR packages. The subsections that follow provide information on each type.

Importing a Gradle Project

Importing a Gradle project means importing a project that already has Gradle build scripts inside. This can be either a Java project or another Android module already developed with Android Studio.

It is relatively easy to import a Gradle project. When Import Gradle Project is selected from the New Module window shown in Figure 7-1, a wizard opens in which you can select the Gradle project directory to be imported.

If you select the folder where the Gradle project is stored, the Module name text box shown in Figure 7-32 will appear so you can provide a unique name for the external Gradle project module.

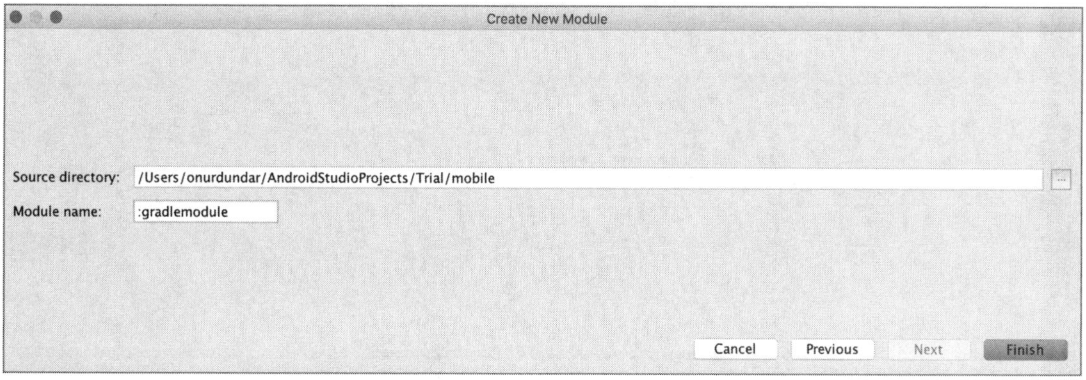

FIGURE 7-32: Gradle project module naming

Android Studio automatically imports the project in the background. When that's done, you can run or debug the imported module.

The method used for running or debugging imported modules depends on the type of module. If it is an Android, Wear, or TV module, it can be run as the examples were in previous sections in this chapter.

When Gradle sync finishes successfully, a new module will appear in Android Project view with the name you gave it in the Create New Module window (see Figure 7-32). As you saw with the previous examples in this chapter, when you select Build APK from the Build menu, an APK file will be generated in the module's `build/outputs/apk` folder.

> **WARNING** *If you are importing an Android module that has an Android SDK version that is not installed on your machine, you should install it to successfully build the module.*

Importing an Eclipse ADT Project

Importing an Eclipse ADT project helps developers migrate Android applications previously developed in Eclipse with ADT. When you select Import Eclipse ADT Project from the module selection window shown in Figure 7-1, the import wizard will ask for the path of the Eclipse ADT project.

> **WARNING** *If your Eclipse ADT project folder isn't missing the* `AndroidManifest.xml` *file or the* `src/`, `res/`, `.project`, *or* `.classpath` *folders, Gradle sync should work correctly. If any of these folders is missing, opening the project in Eclipse would create the missing items for you.*

Select the project path by clicking the button to the right of the Source directory text box. Another text field will become active so you can name the module for your project. In Figure 7-33, we imported a previously implemented application named smartHome.

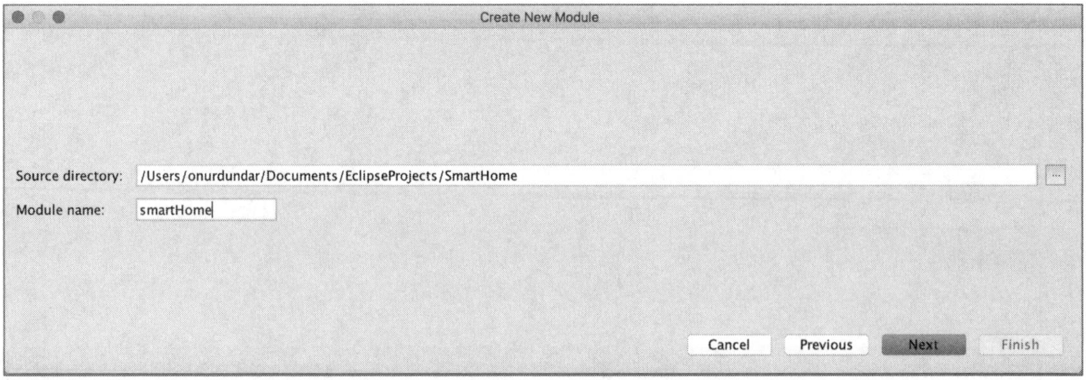

FIGURE 7-33: Module naming for an imported Eclipse ADT project

Next, you need to open the window shown in Figure 7-34. The import wizard will ask you to confirm replacing jar and library dependencies, and Gradle module creation.

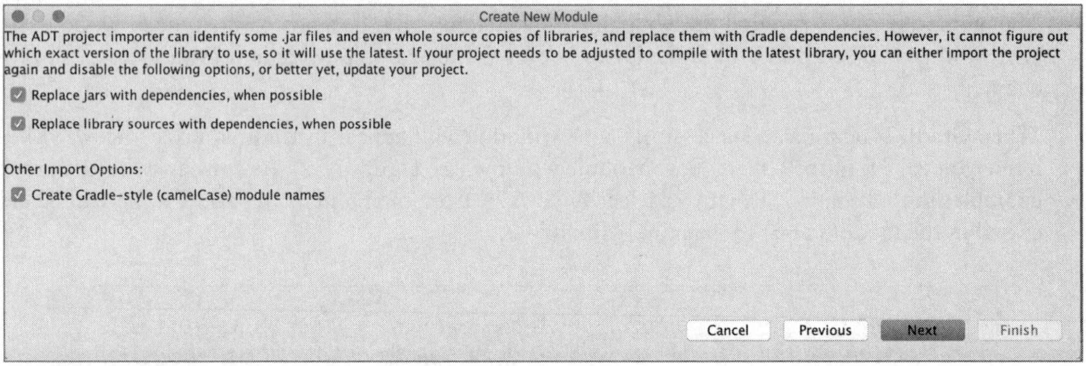

FIGURE 7-34: Dependency replacement confirmation for an imported Eclipse ADT project

If the Eclipse project's Android SDK version is not installed, the setup wizard will prompt you to install a corresponding version from the Android SDK Manager, as shown in Figure 7-35. You can still continue to add the module, but Gradle sync will raise an error to make you install the indicated Android SDK version.

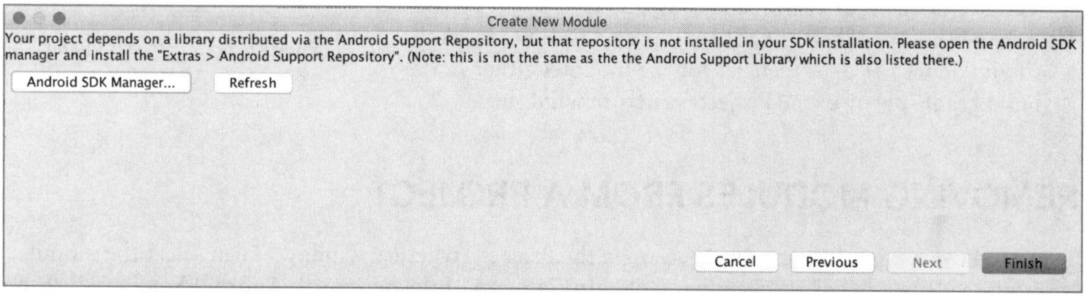

FIGURE 7-35: Missing Android SDK version warning

If the process is successful, you will see the Eclipse application with the name you gave it as a new module in the Project view.

The imported Eclipse ADT project is run as in the previous examples in this chapter. You just need to select the correct device to run or debug the module.

After the Eclipse ADT project has been imported as a module into your Android project, use the Build APK option to generate the module's APK file in the module's new folder, under the Android Studio project's root folder, such as `ProjectRootFolder/smarthome/build/outputs/apk`.

Importing a JAR/AAR Package

Importing a JAR or ARR package is done by including an external Android or Java library in your existing project. JAR files are legacy Java library containers. When you import a JAR file, you are also importing the library in your project. An AAR package is an Android library package, which contains a compressed Android Library module to load into your project.

When you select Import .JAR/.AAR Package from the window shown in Figure 7-1, the window shown in Figure 7-36 opens. In the File name box, point to the `.jar` or `.aar` file to import into your project. In the Subproject name box, enter a name for the module for the Android project.

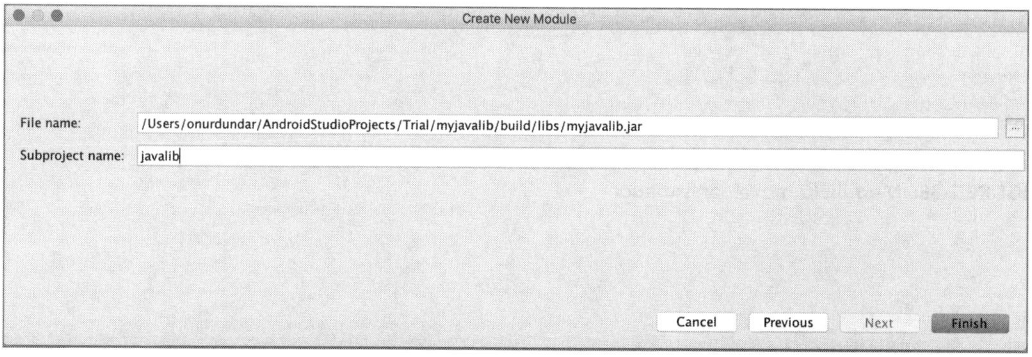

FIGURE 7-36: Importing a JAR/AAR package

As shown in Figure 7-37, after successful Gradle sync, the
Android Project view shows the packages and their build.gradle
files. Android Project view shows only library packages, so there
won't be any access to individual files.

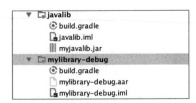

You use these libraries the same way you use the Java and
Android libraries that you added in previous sections. You
should define the dependencies for the modules either in the
build.gradle file or in the Project Structure window.

FIGURE 7-37: Project view of an
imported JAR/AAR module

REMOVING MODULES FROM A PROJECT

To remove a module from a project, open the Project Structure window. Then select the module
from the list and click on the minus (–) button on top of the left panel. Android Studio will then ask
for confirmation, as shown in Figure 7-38.

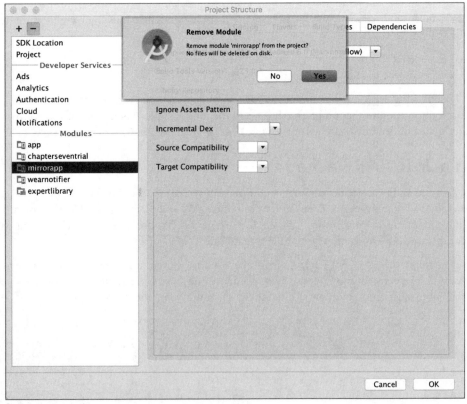

FIGURE 7-38: Module Remove confirmation

Following your confirmation, Gradle sync will update the project and clean up the related files and dependencies.

> **WARNING** *Removing an independent module is easy, but if you are removing a module that is referenced in different modules, such as a library module, you should be careful to identify and appropriately handle the use of the deleted module in other modules.*

SUMMARY

This chapter covered the details of existing modules in an Android Studio project. We discussed in detail the processes for adding new Android Phone & Tablet, Wear, Library, Glass, Auto, and TV modules to a project (including building, running, and debugging configurations). The chapter explored the procedures for releasing APK and JAR/AAR files, and identifying their locations in the project folder.

Next, the chapter covered importing external modules into an existing project and discussed the Google Cloud and Java Library modules.

An understanding of the structure of modules and how they are managed in Android Studio will enable you to better approach the development of complex Android projects with multiple modules. We believe you will benefit from understanding when the modules discussed in this chapter are needed and how they are managed within the project build system.

8

Debugging and Testing

WROX.COM CODE DOWNLOADS FOR THIS CHAPTER

The wrox.com code download for this chapter is found at www.wrox.com/go/expertandroid on the Download Code tab. The code for this chapter is in the Chapter8.zip file.

This chapter covers debugging and testing Android projects using Android Studio and SDK tools. You will use applications from Google's Android samples because those provide excellent use cases for debugging and testing.

Debugging helps to detect flaws and solve possible problems with your software and design. One tricky part of debugging Android applications is that you need to remotely debug a running virtual machine or Android device, which requires a connection to send and receive data from the remote device to your development device. That connection is handled with Android SDK tools.

Android Studio provides a good visual debug console and tools to monitor running applications on the device. Android Studio tools are not yet robust enough for detailed debugging and testing, so this chapter also investigates some of the core tools of Android SDK so that readers can better understand the available debugging options.

In addition to debugging, you need to ensure the application is solid, doesn't crash, and does all required tasks without any problems, so defining all the possible actions as test packages for code and the user interface is a good practice to make a great application.

> **NOTE** *If you have not enabled your Android device for debugging, you should do so first by navigating to Settings ⇨ About Phone (or Tablet, TV) and tap Build Number seven times to unhide the Developer settings. Then navigate to Settings ⇨ Developer Settings to enable USB debugging.*

DEBUGGING ANDROID CODE

For this section, import the Universal Music Player sample application. We'll use that to look at debugging Android code with Android Studio. When you have loaded the project into Android Studio, you are ready to debug the application both on hardware and on a virtual device.

As mentioned in this chapter's introduction, remote connections to Android devices are needed to debug applications remotely. To communicate with a remote Android device, the Android adb (Android Debug Bridge) tool is used. It is delivered with Android SDK and is integrated with Android Studio.

Before proceeding, let's take a look at the details of adb to understand the underlying technology for remote debugging.

Android Debug Bridge

adb (Android Debug Bridge) is a command-line tool that provides communication between a development machine and an Android device. It is delivered with Android SDK and is installed when you install the Android platform-tools package from Android SDK Manager. The Android platform-tools and executable can be found by following the `<sdk-path>/platform-tools/` Android SDK installation path.

There are long-running adb services on both the development host and target Android devices to establish communication.

➤ **adb-server**—adb-server runs on your development machine. It provides the communication between adb and adb-daemon running on the target device.

➤ **adb-daemon**—adb-daemon runs on target Android devices to respond to incoming debug connections from remote host machines.

adb communications happen through TCP ports. When an adb client starts, it should connect with adb-server. If adb-server is not running, the client initiates adb-server to enable connections with the remote device.

adb is at the heart of Android application development. When you run an application, it is installed using adb. If you want to install an Android APK manually, you can run the following command from the terminal if at least one device is already connected to adb-server.

```
adb install <path_to_project>/app/build/outputs/apk/app.apk
```

If more than one device is connected to adb-server, you should modify the command to select a device. The following command lists the connected devices with their unique serial number.

```
adb devices
List of devices attached
emulator-5554    device
f2f6c6c5    device
```

The output shows connected devices so now you can select an emulator to install your apk with the following command. -s instructs adb to select the specified device.

```
adb -s emulator-5554 install <path_to_project>/app/build/outputs/apk/app.apk
```

There are many other commands and options with adb for connecting to Android devices' shells, listing files, transferring files between host and target devices, and so on. Basically, adb works with all IDEs to connect with remote Android devices.

> **REFERENCE** *More information about adb commands can be found at* http://developer.android.com/tools/help/adb.html.

The next section discusses how to use adb to wirelessly debug devices.

Wireless Debugging

The preceding section discussed adb and debugging Android code on remote devices, which can be either an Android emulator or a USB-connected Android device. Wireless debugging can be performed if both devices are in the same local network and can directly connect from a TCP/IP port. However, the devices need to be set up to enable wireless debugging. Therefore, before unplugging your device, run the following adb command after you navigate to the <sdk-path>/platform-tools folder.

```
$ adb tcpip 5555
```

If your emulator is also running, get the device's unique ID from the output of the following command, and enter it in the command on the last line.

```
$ adb devices -l
List of devices attached
f2f6c6c5                device usb:336592896X product:gm4g
model:General_Mobile_4G_Dual device:gm4g_sprout
emulator-5554           device product:sdk_google_phone_x86
model:Android_SDK_built_for_x86 device:generic_x86
$ ./adb -s <device-id> tcpip 5555
```

The device will reconnect with the configured port, which is enabled to allow a wireless debug connection.

Now, let's detach the USB cord from the host machine and find your device's IP address from Settings ⇨ About Phone (or About Tablet, TV) Status. In this example it is 192.168.1.37.

Next, connect adb to our phone wirelessly with following adb command.

```
$ ./adb connect 192.168.1.37
connected to 192.168.1.37:5555
$ ./adb devices -l
List of devices attached
192.168.1.37:5555       device product:gm4g model:General_Mobile_4G_Dual
device:gm4g_sprout
emulator-5554           device product:sdk_google_phone_x86
model:Android_SDK_built_for_x86 device:generic_x86
```

You made it. The wireless connection is successful and you can see your phone on the list of devices. Now, you are ready to run and debug applications using a WiFi connection. When you select Run or Debug from Android Studio, your device will be listed again.

Start Debugging

Now that you understand how the underlying Android debugging mechanism works, you can start debugging with Android Studio using the Universal Music Player sample application.

> **NOTE** *If you have enabled Android Studio ADB Integration, adb should work with Android Studio. However, it is always good to double check adb integration from the Tools menu's Android option, as shown in Figure 8-1.*

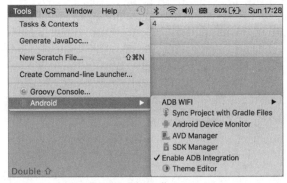

FIGURE 8-1: Android Studio adb integration

Let's insert some break points, starting with the application's entry point, the onCreate function. This will make it easier to explain the steps needed for debugging.

The UniversalMusicPlayer's launcher activity is in the `MusicPlayerActivity.java` file under the `java/com.example.android.uamp/ui` folder (you can locate the activity declaration in the Android Manifest file). We inserted a break point in the onCreate function by double-clicking on the left pane just next to code text as shown in Figure 8-2, but you can simply press Command+F8 on Mac or Ctrl+F8 on Windows.

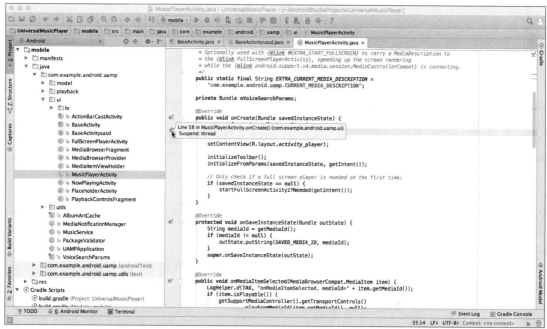

FIGURE 8-2: Break point in Android Studio

At this point, connect your device to a development machine. To quickly start debugging, press Control+d on Mac or Shift+F9 on Windows, or select Debug from the Run menu. Then select the device you will use to debug the application from the Select Deployment window that opens.

When debugging starts successfully, the Debug window is enabled and it displays the progress of the apk installation on the device, as shown in Figure 8-3.

```
Debug    mobile
Debugger  Console

Target device: general_mobile-general_mobile_4g_dual-f2f6c6c5
Installing APK: /Users/onurdundar/AndroidStudioProjects/UniversalMusicPlayer/mobile/build/outputs/apk/mobile-debug.apk
Uploading file to: /data/local/tmp/com.example.android.uamp
Installing com.example.android.uamp
DEVICE SHELL COMMAND: pm install -r "/data/local/tmp/com.example.android.uamp"
    pkg: /data/local/tmp/com.example.android.uamp
Success

Launching application: com.example.android.uamp/com.example.android.uamp.ui.MusicPlayerActivity.
DEVICE SHELL COMMAND: am start -D -n "com.example.android.uamp/com.example.android.uamp.ui.MusicPlayerActivity" -a android.intent.action.MAIN -c android.intent
.category.LAUNCHER
Client not ready yet..
Client not ready yet..
Client not ready yet..
Client not ready yet..
Client not ready yet..
Client not ready yet..
Starting: Intent { act=android.intent.action.MAIN cat=[android.intent.category.LAUNCHER] cmp=com.example.android.uamp/.ui.MusicPlayerActivity }
Connected to process 13728 on device general_mobile-general_mobile_4g_dual-f2f6c6c5
Connected to the target VM, address: 'localhost:8609', transport: 'socket'

5: Debug    TODO    6: Android Monitor    Terminal    0: Messages        Event Log    Gradle Console
```

FIGURE 8-3: Debug window console

Let's dig a little more into the available buttons in the Debug window. First, let's investigate the buttons on the left pane of the Debug window, shown in Figure 8-4.

The first three buttons are used, respectively, to Resume, Pause, and Stop actions for the currently running application.

➤ **Resume**—This button runs the paused application until it reaches a breakpoint or the Pause button is clicked.

➤ **Pause**—This button pauses the running application. The cursor goes to the paused line of the code.

➤ **Stop**—This button stops the debugging process and kills the running application process.

The next two buttons are used for breakpoint operations.

FIGURE 8-4: Debug window left pane actions

➤ **View Breakpoints**—This button shows the list of breakpoints, as shown in Figure 8-5.

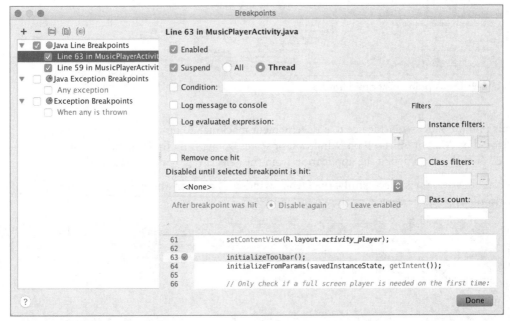

FIGURE 8-5: Breakpoints window

The Breakpoints window enables you to view and configure the breakpoints for detailed use during the debug.

➤ **Mute Breakpoints**—Click this button to disable (but not remove) all active breakpoints.

The sixth button dumps the threads' data, as shown in Figure 8-6.

FIGURE 8-6: Thread dump data

> ➤ **Get thread dump**—A thread dump shows the running threads and stack trace to show the callbacks and current state of the threads for the running application. This view enables you to optimize threads and investigate details of the threads used in your application.

The next two buttons are used to restore the user interface layout to its original state if you have changed some values during debugging.

> ➤ **Restore Layout**—Restores changes back to original state if you have made any changes to layout during debugging.

> ➤ **Settings**—Shows options to show values of variables on code text file, method return values, sort variables in the Frames section alphabetically.

The last three buttons are for settings and window actions to pin, close, and open the help window.

> ➤ **Pin Tab**—Pins debug tool window tab on Android Studio.

> ➤ **Close**—Closes debug window.

> ➤ **Help**—Opens help window.

The action buttons on the top of the debug window (see Figure 8-7) give us the power for step-by-step debugging. For efficiency it is a good idea to make shortcuts of these actions; otherwise, it can be painful to debug applications.

FIGURE 8-7: Debug window actions toolbar

> ➤ **Show Execution Point**—The first button from the left takes you to the current execution point. The shortcut for this action is Option+F10 on Mac and Alt+F10 on Windows.

> ➤ **Step Over**—The second button from the left debugs the application step by step starting from the current position of the cursor in the code. The shortcut is F8 on Mac and Windows. Because Step Over is used repetitively, using your keyboard helps to quickly debug the application and skip lines.

In this section, we started to debug with the onCreate method's first line. To continue debugging, click the Step Over button to step one line. Notice that the cursor advances one line and the action on the line is completed.

➤ **Step Into**—Use the third button from the left to step into a method while debugging the application. The shortcut for the Step Into action is F7 on Mac and Windows.

Step Into allows you to get into the method on the current line of the Java code. For example, when the cursor is on the line LogHelper.d(TAG, "Activity onCreate");, pressing the Step Over button completes the action and continues to the next line. But if you click the Step Into button, you call the LogHelper.d function and continue debugging in that function.

➤ **Force Step Into**—The next button disables any stepping filters and puts you in the function you want to get in. The key combination for this action is Option+Shift+F7 on Mac and Shift+Alt+F7 on Windows.

➤ **Step Out**—The Step Out button takes you out of the method you are currently in during the debugging.

To run the Step Out action, you can press Shift+F8 on Mac and Windows.

➤ **Run to Cursor**—This button makes the debugger work from its current execution point to the current location of the cursor. This is useful if you have skipped through a section of code but decide to go back and debug that section. Click on the line you want to go to and press the Run to Cursor button to debug the code from that line through to the current location of the cursor.

You can perform this action using Option+F9 on Mac or Alt+F9 on Windows.

➤ **Drop Frame**—This takes you back to the method that made the call for the current method. When this button is used, it drops loaded method frames from the stack.

➤ **Evaluate Expression**—The Evaluate Expression action allows you to immediately perform actions or expressions written in the Java code to evaluate them according to the current context. For example, you can evaluate math expression values used for the user interface, which you may not be able to test during development.

It is also possible to test a code fragment because most objects' values would be defined in the debug context so you would get a result for the object (see Figure 8-8). In Figure 8-8, a Boolean expression has been evaluated and the result has been inspected.

You can also open the Evaluate Expression window with Option+F8 on Mac or Alt+F8 on Windows.

Expression evaluation is a powerful tool to quickly test a piece of running code to see whether the output returns the expected result. If it is not returning the assumed value, this helps you determine the problem with variables and understand what you need to work on. This is useful for analyzing sensor values or data read from a server and so on.

To see a result after expression evaluation, make sure you are running an expression that has a return value. If you run a void function, you will see an undefined result, as shown in Figure 8-9. However, because a Boolean expression is used, the result is true.

FIGURE 8-8: Evaluate Expression window

FIGURE 8-9: Undefined result in the Evaluate Expression window

Using the actions available on the Actions toolbar described previously, you can debug your application by running it step by step to investigate what happens at each step. It is also possible to access some of the actions by right-clicking inside the Java class during debugging.

As shown in Figure 8-10, Evaluate Expression, Run to Cursor, and Force Run to Cursor are enabled during debugging. Because these actions are associated with the line of the code and the written expression in the code, only these items can be accessed by right-clicking in the code.

However, if you don't see the details of the threads, navigate to the Debugger tab in the Debug window, as shown in Figure 8-11.

You can see three panes in this window: Frames, Variables, and Watches.

Copy Reference	⌥⇧⌘C
Paste	⌘V
Paste from History…	⇧⌘V
Paste Simple	⌥⇧⌘V
Column Selection Mode	⇧⌘8
Find Usages	⌥F7
Refactor	▶
Folding	▶
Analyze	▶
Go To	▶
Generate…	⌘N
Evaluate Expression…	⌥F8
Run to Cursor	⌥F9
Force Run to Cursor	⌥⌘F9
Add to Watches	
Local History	▶
Compare with Clipboard	
File Encoding	
Create Gist…	

FIGURE 8-10: Actions accessible inside Java code

➤ **Frames**—This pane lists the threads of the Android application. It shows the thread's call stack, and if you click the stack element, you see the thread's variables listed in the Variables pane. Seeing a thread's call stack helps you understand the steps, and you can look over the variables and objects in the threads for a better understanding of how the application runs.

You can switch between threads by clicking on the drop-down box on top of the pane.

➤ **Variables**—This pane shows the list of variables and their values at the current state of the selected stack element from the Frames pane. In this pane, you can manipulate the variables. You can see a list of the actions that you can perform by right-clicking a variable, as shown in Figure 8-12.

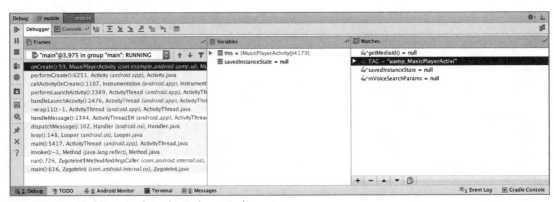

FIGURE 8-11: Debugger tab in the Debug window

➤ **Watches**—This pane lists the variables you've manually selected to watch over the debugging process.

Figure 8-13 shows that in the current state, just at the beginning of the onCreate function, only the TAG variable is defined; the rest are null. null is a constant and static variable, so its value is assigned already. For this example, some of the variables defined in the MusicPlayerActivity.java class have been randomly inserted.

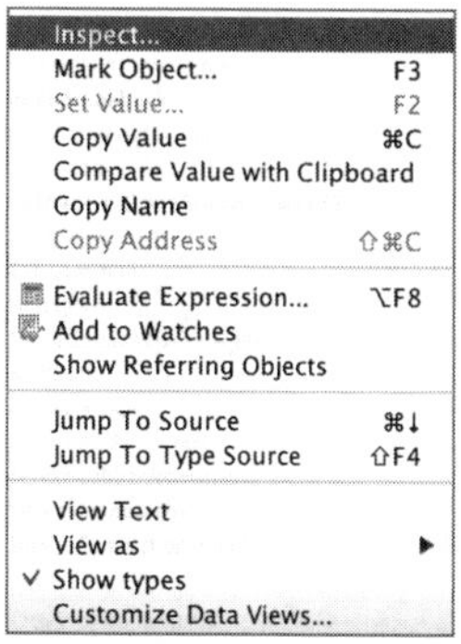

FIGURE 8-12: Variable actions

FIGURE 8-13: Watches pane

The Watches property enables you to group a number of variables during debugging to see their values change during execution. That way, there's no need to search through the files to try to catch their values each time. When there has been a change to the variable, you can see it in the Watches window.

You can add variables to watch by right-clicking and selecting Add to Watches or you can select the variable and drag it to the window. Another way is to use the + button on the pane, shown at the bottom of Figure 8-13.

Finally, we should mention that Android Studio tries to help as much as it can to provide all the information about variables and values during debugging. When you step over any line, it highlights the variable and object values instantly, as shown in Figure 8-14.

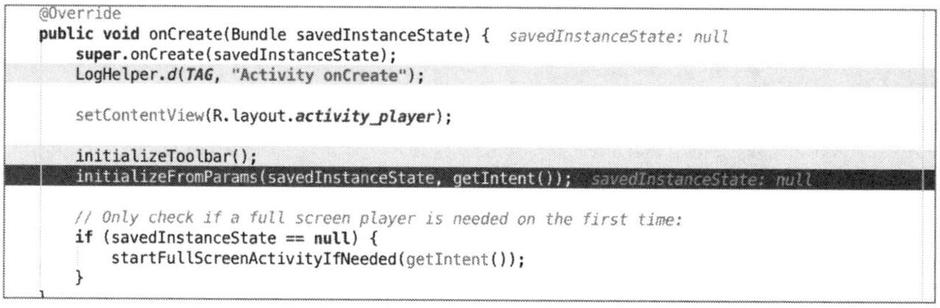

```
@Override
public void onCreate(Bundle savedInstanceState) {   savedInstanceState: null
    super.onCreate(savedInstanceState);
    LogHelper.d(TAG, "Activity onCreate");

    setContentView(R.layout.activity_player);

    initializeToolbar();
    initializeFromParams(savedInstanceState, getIntent());   savedInstanceState: null

    // Only check if a full screen player is needed on the first time:
    if (savedInstanceState == null) {
        startFullScreenActivityIfNeeded(getIntent());
    }
}
```

FIGURE 8-14: Highlighted variables and values in code

It is helpful to see light green highlights on the debug line instead of needing to hover over the variable, as shown in Figure 8-15. The highlighting saves significant time and practically debugs the application for you.

```
@Override
public void onCreate(Bundle savedInstanceState) {    savedInstanceState: null
    super.onCreate(savedInstanceState);
    LogHelper.d(TAG, "Activity onCreate");

    setContentView(R.layout.activity_player);

    initializeToolbar();
    initializeFromParams(savedInstanceState, getIntent());    savedInstanceState: null

    // Only check  savedInstanceState = null  needed on the first time:
    if (savedInsta
        startFullScreenActivityIfNeeded(getIntent());
    }
}
```

FIGURE 8-15: Hovering over a variable to reveal its value

This section covered most of the tools and shortcuts required to efficiently debug an Android application. During the debugging, you always need to watch the stack and threads to see what is changing and whether you're getting the expected behavior or value of the objects. In the next section, you learn to use Android Monitor to trace memory, CPU, and GPU usage, as well as network activity.

ANDROID MONITOR

Android Monitor should be one of your best friends while developing your application in Android Studio because it includes the useful debugging monitoring tools logcat, Memory, CPU, GPU, and Network. The tabs for these monitors are available in the Android Monitor window (see Figure 8-16).

> **NOTE** *You don't need to be in debug mode to use tools in Android Monitor, as the goal is to optimize the release builds.*

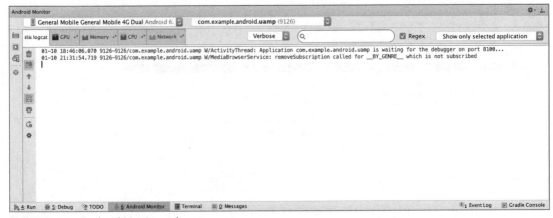

FIGURE 8-16: Android Monitor tabs

Before we cover the detailed use of these monitoring tools, let's look at the common actions you can perform in Android Monitor.

Figure 8-17 shows the three buttons available near the top left of the Android Monitor window. These are used to capture and dump data from an Android device. You can take screenshots with the first button and record screen activity with the second. With the last button, you can capture to a .txt file the system information listed in Figure 8-17 .

FIGURE 8-17: List of system information available with Android Monitor

In addition to these information-capturing buttons, there's a red button that is used to terminate selected processes. The following list describes in detail how to use those actions:

➤ **Screenshot**—This button is used to capture the current screen from the connected device during debugging. Right after you click the Screenshot button, a dialog box appears that shows the progress of capturing and transferring the image to the host machine. Then the window shown in Figure 8-18 displays a preview of the screenshot. Here you can perform minor editing with the tools provided at the top of the window.

FIGURE 8-18: Screenshot preview window

You can save screenshots to your development machine with the Save button. To recapture a screenshot, click the Recapture button. If you want to view the screenshot as it would appear on a phone, check the Frame Screenshot option before saving.

➤ **Screen Recording**—This button records a video of the attached Android device's screen to your host machine. When you first click the button, a dialog box appears in which you configure the bitrate and resolution (height and width of video) options, as shown in Figure 8-19.

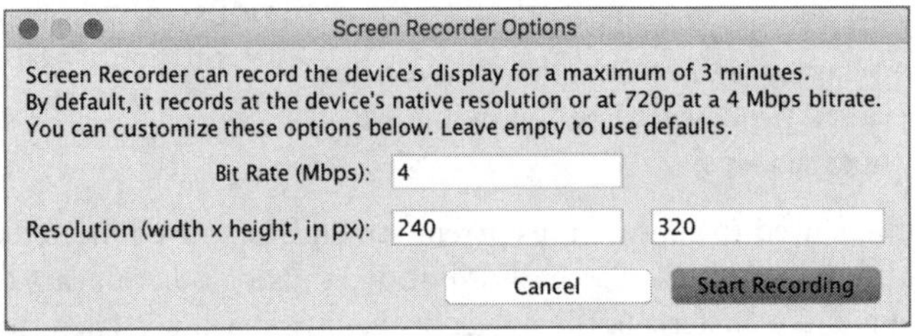

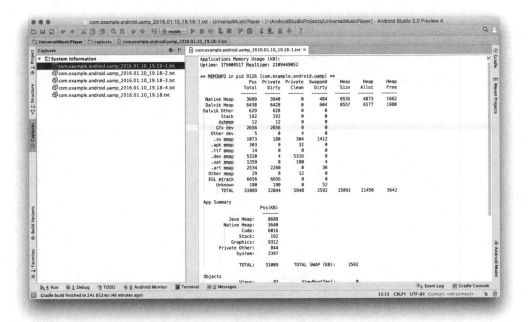

FIGURE 8-19: Screen Recorder Options dialog box

You can share a screen recording with your team, which is useful when you want to investigate and discuss the behavior of the user interface.

➤ **Capture System Information**—Click this button to get the system information shown in Figure 8-17. When you click any item from the list, the command is sent to adb and you get the corresponding data as text. After the data is received, it will be opened and saved in the `Captures\System Information` folder under the module's path. The result is that you can see all the captured system information in the Captures window, as shown in Figure 8-20.

FIGURE 8-20: System information Captures window

All the capturing tools are important if you want to share your application's data with your teams and peers to help them understand any problems there might be. You can also use this information to compare different devices' system information so that you can understand what can be improved.

For example, the dump (system information after the capture action) shown in Figure 8-20 is a device's memory usage information provided in a text file. With this information, you can investigate memory usage by system processes.

You are also able to select the connected devices and available processes from lists above the Android Monitor window, as shown in Figure 8-21. You can see the two connected devices: a smartphone and a running emulator. Capture outputs will give the selected device's information or screenshot.

FIGURE 8-21: Device and process selection

The following sections investigate the other tools in Android Monitor.

Using logcat

Logcat is one of the most useful tools for Android application development because Android applications do not run on a shell with stdin/stderr. It is possible to print output to text files during debugging or running but it is expensive to implement. A better solution is the Android Log class and logcat functionality for all applications.

There is an absolute need to follow up with the running code for tracing and getting logs to analyze code and get exception messages and other warnings, errors, and information.

If you select No Filter instead of Show only selected applications from the list at the top right of the logcat window, you will see all the log messages from the system and other running applications and services, as shown in Figure 8-22.

```
logcat  GPU  Memory  CPU  Network         Verbose        Q              Regex    No Filters
  01-10 21:58:10.486 364-364/? I/VM_BMS: power_supply_update: ocv_uv=4305072 ibatt=170899 soc=99 seq_num=27054
  01-10 21:58:25.691 272-2136/? D/MDnsDS: MDnsSdListener::Monitor poll timed out
  01-10 21:58:25.691 272-2136/? D/MDnsDS: Going to poll with pollCount 1
  01-10 21:58:58.711 364-364/? I/VM_BMS: power_supply_update: ocv_uv=4303959 ibatt=177435 soc=98 seq_num=27055
  01-10 21:59:28.675 272-2131/? E/NetlinkEvent: NetlinkEvent::FindParam(): Parameter 'UID' not found
  01-10 21:59:33.682 4119-13022/? D/NetlinkSocketObserver: NeighborEvent{elapsedMs=2119120952, 192.168.1.1, [A0E4CB1D65B9], RTM_NEWNEIGH, NUD_REACHABLE}
  01-10 21:59:37.962 4119-13022/? D/NetlinkSocketObserver: NeighborEvent{elapsedMs=2119125232, 192.168.1.1, [A0E4CB1D65B9], RTM_NEWNEIGH, NUD_STALE}
  01-10 21:59:43.862 272-2131/? E/NetlinkEvent: NetlinkEvent::FindParam(): Parameter 'UID' not found
  01-10 21:59:47.116 364-364/? I/VM_BMS: power_supply_update: ocv_uv=4302871 ibatt=173763 soc=98 seq_num=27056
  01-10 22:00:35.369 364-364/? I/VM_BMS: power_supply_update: ocv_uv=4301773 ibatt=173462 soc=98 seq_num=27057
  01-10 22:01:23.558 364-364/? I/VM_BMS: power_supply_update: ocv_uv=4300697 ibatt=164282 soc=98 seq_num=27058
  01-10 22:02:11.743 364-364/? I/VM_BMS: power_supply_update: ocv_uv=4299624 ibatt=169178 soc=98 seq_num=27059
  01-10 22:02:59.954 364-364/? I/VM_BMS: power_supply_update: ocv_uv=4298591 ibatt=147783 soc=98 seq_num=27060
  01-10 22:03:48.152 364-364/? I/VM_BMS: power_supply_update: ocv_uv=4297583 ibatt=160043 soc=98 seq_num=27061
  01-10 22:04:36.345 364-364/? I/VM_BMS: power_supply_update: ocv_uv=4296542 ibatt=164604 soc=98 seq_num=27062
  01-10 22:05:12.242 4119-4128/? I/art: Background partial concurrent mark sweep GC freed 72502(4MB) AllocSpace objects, 44(1064KB) LOS objects, 33% free, 18MB/27MB,
      paused 2.156ms total 141.161ms
  01-10 22:05:24.653 364-364/? I/VM_BMS: power_supply_update: ocv_uv=4295538 ibatt=159613 soc=98 seq_num=27063
  01-10 22:06:12.834 364-364/? I/VM_BMS: power_supply_update: ocv_uv=4294536 ibatt=161770 soc=98 seq_num=27064
  01-10 22:07:01.017 364-364/? I/VM_BMS: power_supply_update: ocv_uv=4293559 ibatt=153214 soc=98 seq_num=27065
  01-10 22:07:49.210 364-364/? I/VM_BMS: power_supply_update: ocv_uv=4292608 ibatt=149080 soc=98 seq_num=27066
  01-10 22:08:37.574 364-364/? I/VM_BMS: power_supply_update: ocv_uv=4291642 ibatt=154164 soc=97 seq_num=27067
  01-10 22:09:25.789 364-364/? I/VM_BMS: power_supply_update: ocv_uv=4290707 ibatt=149582 soc=97 seq_num=27068
```

FIGURE 8-22: No Filters output in logcat

Log messages are classified as Verbose, Debug, Info, Warn, and Error. The Log class allows you to write messages according to these levels using v, d, i, w, and e functions respectively. Those functions take two parameters. The first parameter is used to define a tag and the second is a string with the message you want to print. The message can be an exception message or any other message you want. Listing 8-1 shows the use of the Log functions.

LISTING 8-1: Log function use

```
//It is better to define a tag in a static constant field
private static final String TAG = "Class Name";
//Verbose is used to print casual messages according
Log.v(TAG, "Starting");
//Debug is used to print messages in debug mode
Log.d(TAG, object.Length.toString());
//Information to print out general system flow messages
Log.i(TAG, "Starting to Execute ...");
//Warning to print out warning messages especially when exceptions catched
during execution
Log.w(TAG, ex.message);
//Error to print execution errors which is most important messages
Log.e(TAG, "class returned null");
```

The five types of Log messages help you to filter messages so that you don't get lost in thousands of messages on the logcat screen. You can use the log type selection box in the middle of the screen, as shown in Figure 8-23.

FIGURE 8-23: Log type selection

You can also use the Search box with or without regex to filter log messages.

Using Memory Monitor

Memory monitor enables you to watch the memory use of your application's threads and collections. This information enables you to analyze memory use. With that information, you can optimize your application's memory use and data collection to increase performance and enhance user experience. Nobody wants an application with heavy memory use or leaks that may impact the system and even cause a crash.

The Memory tab shows a flow chart for the application running on the target device, as shown in Figure 8-24.

When you start using Memory monitor the chart changes to reflect memory usage. In the chart, the vertical (y) axis shows memory in use and the horizontal (x) axis tracks the time. Whenever you perform an action, the memory use changes over the duration of the action. However, it is not enough to analyze memory use just by looking at how it changes over time. You need to go deeper into the details, so you need to dump the Java Heap to track the allocation of memory.

To dump the Java Heap, click the button shown in Figure 8-25.

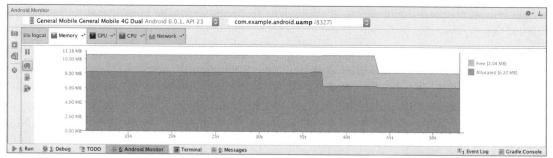

FIGURE 8-24: Memory monitor

The Dump Java Heap button retrieves the threads and the threads' members' current memory allocations and writes them to a `.hprof` file. In order to analyze this information, Android Studio opens the dump file, as shown in Figure 8-26.

FIGURE 8-25: Java Heap dump

Class Name	Tota...	Hea...	Siz...	Shall...	R... ▼
FinalizerReference (java.lang.ref)	881	354	36	12744843908	
byte[]	1796	244	0	4722547225	
Bitmap (android.graphics)	17	11	47	517	464304
MusicService (com.example.android.ua	1	1	106	106	234262
Object[] (java.lang)	2729	643	0	2093623313	
MediaMetadata (android.media)	16	16	16	256	232138
ArrayMap (android.util)	201	128	24	3072	228662
Bundle (android.os)	36	35	23	805	227789
MusicProvider (com.example.android.u	1	1	24	24	227347
ConcurrentHashMap (java.util.concurre	10	3	60	180	226438
ConcurrentHashMap$Node[] (java.util.c	7	2	0	192	226417
ConcurrentHashMap$Node (java.util.co	35	17	24	408	226410
MutableMediaMetadata (com.example..	14	14	16	224	226309
LinkedHashMap (java.util)	33	10	53	530	223275

Instance	Depth	Shallo...	Domi...
▼ ① 0 = {MusicService@315622784 (0x12d00580)	1	106	234262
▶ ① mCarConnectionReceiver = {MusicService$2:	2	20	20
▶ ① mCastConsumer = {MusicService$1@31457	2	12	12
▶ ① mCastManager = {VideoCastManager@314!	1	192	473
① mCurrentIndexOnQueue = 0			
▶ ① mDelayedStopHandler = {MusicService$Del:	2	32	56
① mIsConnectedToCar = false			
▶ ① mMediaNotificationManager = {MediaNotific	2	72	51024
▶ ① mMediaRouter = {MediaRouter@31457947	2	16	84
▶ ① mMusicProvider = {MusicProvider@315603(	2	24	227347
▶ ① mPackageValidator = {PackageValidator@3:	2	12	15964
▶ ① mPlayback = {LocalPlayback@315524864 (	2	58	249
▶ ① mPlayingQueue = {ArrayList@315603072 ((	2	20	388
① mServiceStarted = false			

Reference Tree

	De...	Shallow...	Domin...
▼ ≡ com.example.android.uamp.MusicService@315622784 (0x12d00580)	1	106	2342621
▶ ≡ 🔗 this$0 in **android.service.media.MediaBrowserService$ServiceBinder@314582720 (0x12c026c0)**	0	28	28
▶ ≡ mOuterContext in android.app.ContextImpl@314650112 (0x12c12e00)	2	121	678
▶ ≡ mCallback, mService in com.example.android.uamp.LocalPlayback@315524864 (0x12ce8700)	2	58	249
▶ ≡ mCallback, mService in com.example.android.uamp.LocalPlayback@315524864 (0x12ce8700)	2	58	249
▶ ≡ mService in com.example.android.uamp.MediaNotificationManager@314706848 (0x12c20ba0)	2	72	51024
▶ ≡ this$0 in com.example.android.uamp.MusicService$1@314579456 (0x12c01a00)	2	12	12
▶ ≡ this$0 in com.example.android.uamp.MusicService$2@315602240 (0x12cfb540)	2	20	20
▶ ≡ mContext in android.media.session.MediaController@315518544 (0x12ce6e50)	3	45	149
▶ ≡ mContext in android.media.session.MediaController@315517968 (0x12ce6c10)	3	45	133
▶ ≡ mContext in android.net.wifi.WifiManager@315603584 (0x12cfba80)	4	28	72
▶ ≡ this$0 in com.example.android.uamp.MusicService$MediaSessionCallback@314579600 (0x12c01a90)	4	16	16
▶ ≡ mContext in android.app.LoadedApk$ReceiverDispatcher@315517920 (0x12ce6be0)	4	38	210
▶ ≡ mBase in android.view.ContextThemeWrapper@315602624 (0x12cfb6c0)	4	32	32

FIGURE 8-26: hprof memory dump

The `.hprof` file displays a really detailed memory allocation output for your application and system. In the Class Name pane, you can navigate to a Java class's objects and check the Instance pane to see how its members use memory. The Reference Tree pane shows the hierarchical view of objects and their values.

The Java Heap dump shows the instantaneous allocation of the application's Java Heap. However, you may want to observe and investigate how your application allocates memory while it's running. To track memory allocation this way, click the button below the Dump Java Heap button.

When you start allocation tracking, you see an icon on the memory timeline where it starts recording. After you collect enough data for your purposes, click the allocation tracking button again to stop tracking. When tracking finishes Android Studio opens an `.alloc` file that contains the recorded data, as shown in Figure 8-27.

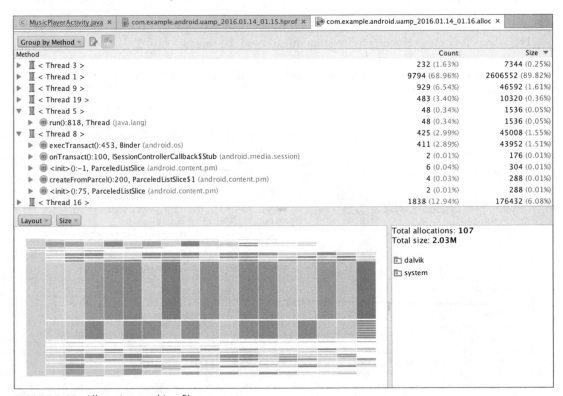

FIGURE 8-27: Allocation tracking file

In the top pane, you see threads and their allocations and count during execution. In the bottom pane, you can group allocators, as shown in Figure 8-28.

In addition to tracking and monitoring, you can also use the button above the Dump Java Heap button to force the Java Garbage Collector to work. Use the top button to pause memory allocation tracking.

FIGURE 8-28: Allocation tracking with allocator grouping

Using CPU Monitor

CPU monitor shows you how much computational power you use during the execution of your application. With this data, you can analyze the running application for further performance enhancements and optimized coding.

Figure 8-29 shows a typical display when you monitor an application. The vertical axis shows the application's CPU usage by percentage.

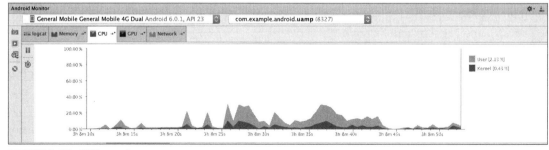

FIGURE 8-29: CPU monitor

There aren't as many actions in CPU monitor as there are in Memory monitor. CPU monitor has only a tracker to get CPU usage by threads into a *trace* file so you can investigate the performance of the application's functions and threads. This is a very detailed file so you should zoom in after it's opened and then navigate to the methods you are looking for to see their CPU usage and execution details as shown in Figure 8-30.

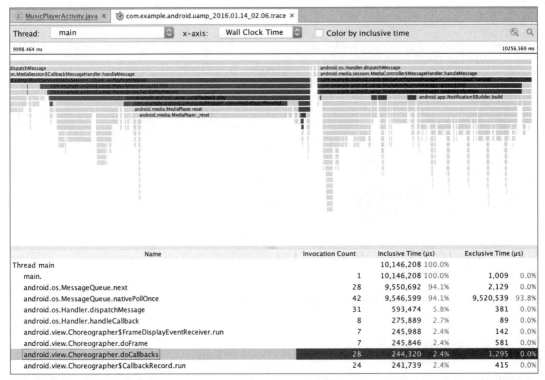

FIGURE 8-30: CPU track dump

CPU monitor helps you find and resolve computational bottlenecks and optimize execution time.

Using GPU Monitor

GPU monitor shows an Android application's GPU use with a list of operations and their instant performance, as shown in Figure 8-31.

The color-coded list at the right of the screen specifies the GPU operations monitored, and the graph is a timeline showing the color-coded operation details. Using this monitoring tool, you can get only basic data about what is happening.

To get detailed GPU tracing, you have to modify your application to activate tracer and dump tracing data, and your device needs to be rooted (that is, given root permissions by unlocking the bootloader and modifying your device). Some other profiling options can also be found on the Android devices' developer options menu at `http://developer.android.com/tools/performance/` `profile-gpu-rendering/index.html`.

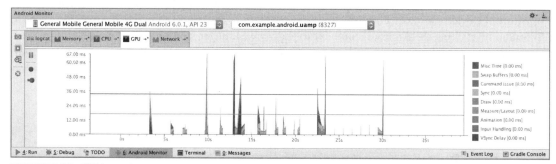

FIGURE 8-31: GPU monitor

GPU debugging and tracing is too large a subject to cover in one section; if you are a game developer and use Open GL heavily, the following resources are the best guides to learn more about GPU performance analysis:

```
http://developer.android.com/tools/help/gltracer.html
http://tools.android.com/tech-docs/gpu-profiler
```

Using Network Monitor

Network monitor is there for you to work on your application's network optimization. This is another crucial point to think about while designing the application; you should consider that any Android application with an Internet connection would eventually be used on 3G/4G network, which is more costly than your home/office network. Even worse, network usage can greatly affect the power consumption of your application. Efficient use of mobile networks will encourage users to run the application without a second thought about consuming a lot of network data and losing battery life.

Network monitor is a basic monitoring tool compared to other monitors because you can observe only the incoming (Rx) and outgoing (Tx) bytes in a timeline. The vertical (y) axis shows the amount of data received or uploaded and changes as the horizontal (x) axis advances. There is no dump tool, only the timeline shown in Figure 8-32.

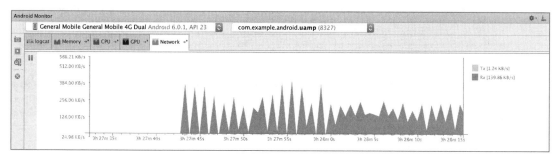

FIGURE 8-32: Network monitor

ANDROID DEVICE MONITOR

In addition to the Android Monitor tools, you have access to a legacy tool from Android SDK, the Android Device Monitor. This tool runs as a separate application and can be started from Android Studio by clicking the Android icon next to the SDK Manager icon on the toolbar. You can also select it from the Tools menu, under the Android section. Right after you click on the icon, Android Device Monitor will run, as shown in Figure 8-33.

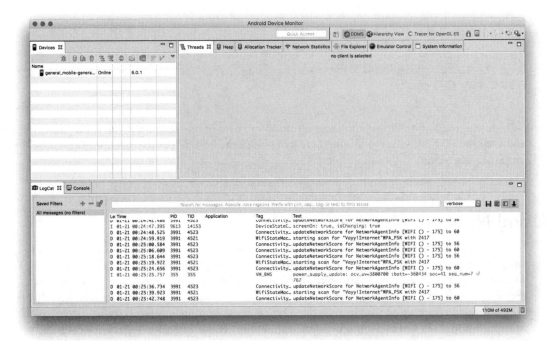

FIGURE 8-33: Android Device Monitor

> **NOTE** *If you start Android Device Monitor from Android Studio, adb will be allocated for Android Device Monitor and Android Studio's connection with the target device will be lost.*

If you were an Eclipse user, you may recognize Android Device Monitor's user interface. It is similar to Eclipse's DDMS perspective view, which was the debug context UI in the Eclipse IDE when debugging Android applications.

Android Device Monitor provides some extra debugging outputs for developers in addition to the tools provided within Android Studio. Following is the list of common monitoring tools in Android Device Monitor with Android Monitor.

➤ **Logcat**—Logcat appears at the bottom of the window in Figure 8-33. Logcat in Android Device Monitor prints the same output as in Android Monitor, but the output format is column-based texts here. The rest of the functions (filtering and searching messages) are the same.

➤ **Devices**—In this tab, you can manage devices, take screenshots, and stop running processes, as you did in Android Monitor. You can also initiate heap and memory dump, and you can see the dumps in tabs at the right of the window. Threads, Heap, Allocation Tracker, and Network Statistics are available as well. Similar to Android Monitor, it is also possible to trace GPU usage.

➤ **Threads**—This tab of Android Device Monitor is similar to CPU Monitor in Android Monitor. You activate thread tracing by clicking Update Threads or Start Method Profiling to get execution data for an Android application's threads. When you click Update Threads, you will see refreshed data about Android threads on the Threads tab, as shown in Figure 8-34.

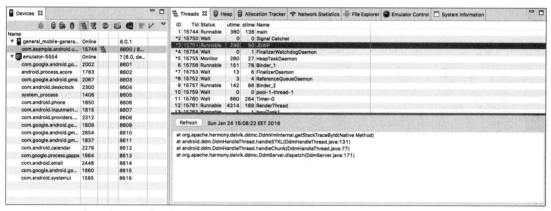

FIGURE 8-34: Threads in Android Device Monitor

If you start tracing an application's threads and functions by clicking Start Method Profiling and then stop tracing after a while, you can get a trace file, as in Android Monitor, to analyze functions and the threads' performance. The .trace file opens automatically, right after you stop tracing, as shown in Figure 8-35.

This tool can be used together with Android Monitor's CPU Monitor, for a little more detailed analysis.

➤ **Heap**—This feature is similar to the Memory Monitor tool in Android Monitor. It is used to analyze an Android application's threads' memory allocations from the Android heap.

You can activate Heap tracing by clicking the Update Heap button on the Devices menu; then you can navigate to the Heap tab and get the threads' current allocations, as shown in Figure 8-36.

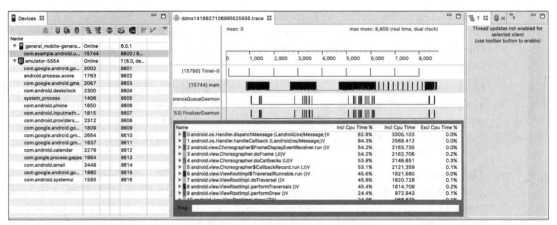

FIGURE 8-35: Thread trace

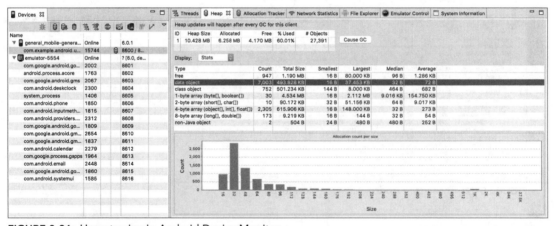

FIGURE 8-36: Heap tracing in Android Device Monitor

You can also dump the hprof file for further analysis of memory allocations using the Dump HPROF File button, and can navigate to the Allocation Tracker tab to instantly monitor the memory allocation of the Android application.

➤ **Network Statistics**—This tool is the same as Network Monitor; you just need to activate network tracking to see RX and TX bytes.

➤ **Emulator Control**—This tab includes options to help with the debugging process within an Android virtual device by mocking SMS, phone calls, and sensors. This feature is not active in Android Studio; instead, it is activated when you launch a virtual device. The next section, "Android Virtual Device Extended Controls," discusses this in more detail.

The following list describes the extensions in Android Device Monitor, which may ease some debugging processes on Android devices.

➤ **UI XML Dump**—This function allows you to dump the XML layout and hierarchy of the currently running application's user interface on the Android device. You can start dumping the XML of the view by clicking the XML dump icon in Device view, as shown in Figure 8-37.

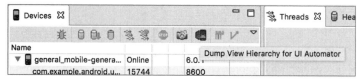

FIGURE 8-37: XML dump icon

As shown in Figure 8-38, when the dump finishes, a new tab opens in the Android Device Monitor. You can click any view to see all the detailed features of the views.

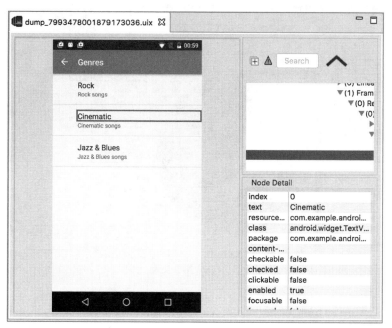

FIGURE 8-38: XML dump view

➤ **File Explorer**—This tool, shown in Figure 8-39, enables you to see files and folders in a tree view. In this view you can transfer files between the Android device and the host machine as well as delete files and create folders using the buttons at the top of view.

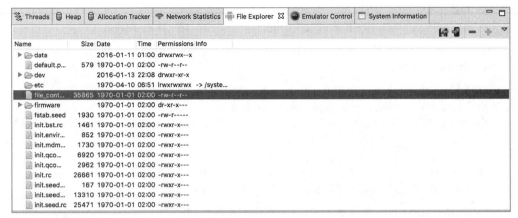

FIGURE 8-39: Android Device Monitor File Explorer

➤ **System Information**—This view uses a pie chart to provide quick information about the device CPU load, Memory usage, and Frame Render Time (see Figure 8-40).

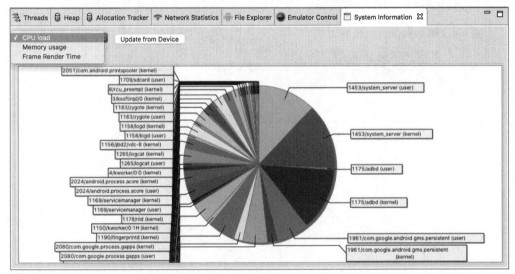

FIGURE 8-40: System Information view in Android Device Monitor

➤ **Hierarchy View**—The tools mentioned previously in this list are part of Android Device Monitor's DDMS (Dalvik Debug Monitor Server). Android Device Monitor has another useful tool to help you analyze your application's user interface layouts and see their render time to optimize the user interface. To display the view hierarchy, click a process from the list and the corresponding hierarchy view will be generated to the right, as shown in Figure 8-41.

The Hierarchy view contains red, green, and yellow circles that indicate how long it takes to render that particular view. Double-click an object to zoom in, as shown in Figure 8-42.

By using Hierarchy view, you can investigate the bottlenecks on your user interface and improve performance for a better user experience.

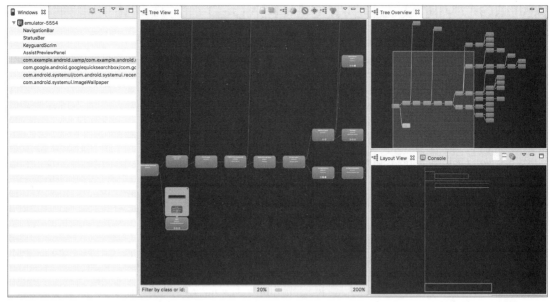

FIGURE 8-41: Hierarchy view generation

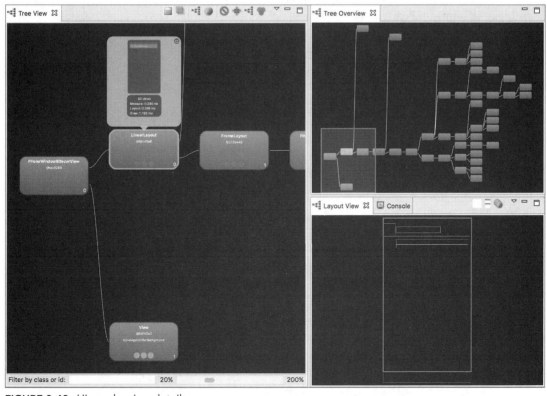

FIGURE 8-42: Hierarchy view details

ANDROID VIRTUAL DEVICE EXTENDED CONTROLS

Some features of Android smart phones and tablets, such as phone calls, handling SMS, GPS data, and fingerprint sensors, might be difficult to virtually emulate on regular personal computers. To make tests easier on virtual devices, Emulator Controller was used by developers in Eclipse. Now, we have extended features embedded into virtual devices that can be accessed from a running virtual device's toolbar by clicking the ellipsis (...) button shown at the bottom of Figure 8-43.

When you click the ellipsis button, the Extended controls window, shown in Figure 8-44, opens.

The extended controls of AVD are as follows:

➤ **Location**—Location helps mock location data on a running Android emulator to check that your application is getting data correctly and your algorithms do the location-related calculations correctly. Trying location simulation is pretty easy. If your AVD is generated by using a Google API image, you can open Google Maps in AVD, set the latitude and longitude, press Send, and press the Locate button in Google Maps, and you are at the location.

FIGURE 8-43: Virtual device toolbar

In the sample, we pointed to the River Thames in London and Google Maps took us there, as shown in Figure 8-45.

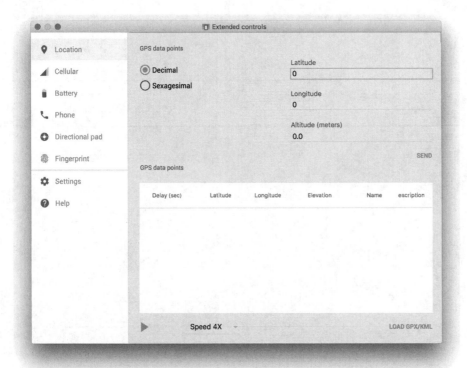

FIGURE 8-44: Extended controls window

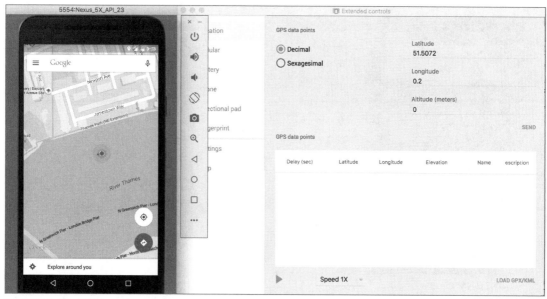

FIGURE 8-45: GPS location setting

However, one static location is not always enough. It is also possible to load a GPX/KML data file to simulate a changing location on an emulator by setting the speed. These buttons and actions can be seen at the bottom of Figure 8-45.

➤ **Cellular**—Cellular allows you to emulate the network status of a phone. You can select EDGE, GPRS, or None as the network type, and voice and data status as either Roaming or Home to simulate different types of network statuses. That enables you to see the behavior of your application. In this way, you can debug and catch unexpected errors.

➤ **Battery**—Battery simulates the battery status on the emulator. The aim is to test how your application works under low and high power conditions.

➤ **Phone**—This feature allows you to call and send SMS to your emulator. If your application works with calls or SMS, you can use this feature for better testing. It is pretty straightforward to call or send SMS. Just type a phone number and press Call or Send Message, and your emulator will show that it's receiving a call from the number, as shown in Figure 8-46.

➤ **Directional Pad**—This feature emulates an Android TV remote controller with directions and playback buttons.

➤ **Fingerprint**—Fingerprint allows you to define a fingerprint template on an emulator and test how it works in your application.

➤ **Help**—This tab, shown in Figure 8-47, includes the keyboard shortcuts to directly access any feature mentioned in this list.

Now, you should be aware of the power of AVD and be able to take advantage of all provided features.

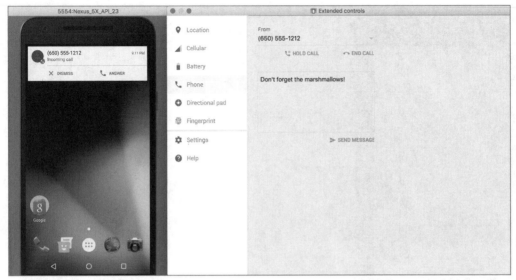

FIGURE 8-46: Emulator call

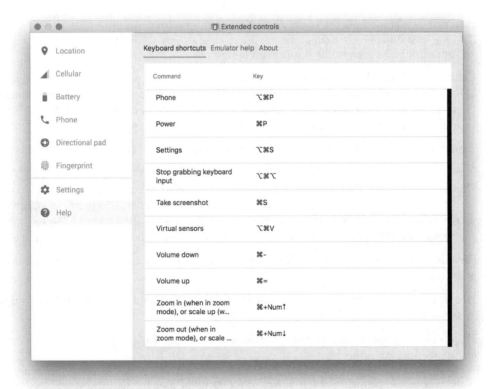

FIGURE 8-47: Android virtual device keyboard shortcuts

USING LINT

Previous sections of this chapter discussed the tools that help with debugging and performance analysis of your Android application. This section visits another tool, Android Lint, which helps you to detect any kind of error, warning, or suggestions for written code.

To run Android Lint, select the Analyze menu and click Inspect Code. Android Lint gives you the opportunity to improve your code, detect possible bugs, correct typos, resolve accessibility of objects, investigate assignment problems, and so on.

It is good practice to run Android Lint before committing code to a version control server when new patches are added to an application.

Figure 8-48 shows example output of Android Lint analysis. The Inspection window shows all findings of Android Lint.

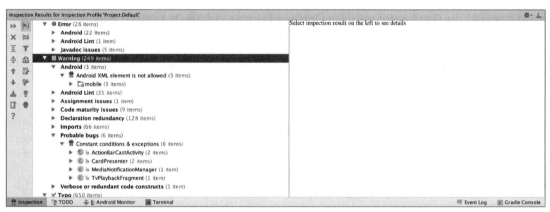

FIGURE 4-48: Android Lint output

Together with Android Studio and Gradle, you can also run Lint during the build and inspect code. In order to set Lint configurations in the Gradle script inside Android, you should add `lintOptions`. In the sample UniversalMusicPlayer's Gradle script, the following section is provided:

```
Android {
...
lintOptions{
        abortOnError true
}
...
}
```

To see other Lint options for Gradle, visit `http://google.github.io/android-gradle-dsl/ current/com.android.build.gradle.internal.dsl.LintOptions.html#com.android.build .gradle.internal.dsl.LintOptions`.

TESTING ANDROID CODE AND APPLICATION UIS

Testing might be the most underestimated topic in the whole software development lifecycle. Every developer and all projects claim to conduct tests, but few conduct the right tests with proper coverage. Tests can help detect bugs and defects before they appear. They greatly help distributed teams to work in harmony without breaking each other's code.

> **NOTE** *Testing is a huge topic that could easily fill an entire book if covered adequately. This chapter cannot dig into every detail, but it does cover common requirements and best approaches to making your projects testable.*

Most test-related APIs in Android have been available since the first "real" release, version 1.5 (Cupcake), and some were available even with the initial release of Android SDK. Yet this didn't make tests appealing to many Android developers. Android Studio, with Gradle, aims to make writing tests and following test-driven development easier. To create tests for the following sections, you will use a pre-written simple example. Download the example code for this chapter, named "beginning," and import the project.

The example is a binary calculator that converts decimal to binary and vice versa via a simple UI. The project has an interface and four class files, consisting of an activity, two fragments, and a class to encapsulate logic. Later, we introduce a few more classes and refactor the current classes. You can run and experiment with the application. Listings 8-2 through 8-6 contain the code for the application. If you are going to create the classes manually instead of downloading them, pay attention to the package names.

LISTING 8-2: MainActivity class code

```
package com.expertandroid.chapter8.binary;

import android.support.design.widget.FloatingActionButton;
import android.support.v7.app.AppCompatActivity;
import android.support.v7.widget.Toolbar;
import android.support.v4.app.Fragment;
import android.support.v4.app.FragmentManager;
import android.support.v4.app.FragmentPagerAdapter;
import android.support.v4.view.ViewPager;
import android.os.Bundle;
import android.view.View;
import com.expertandroid.chapter8.binary.ui.BinaryFragment;
import com.expertandroid.chapter8.binary.ui.CalculatorFragment;
import com.expertandroid.chapter8.binary.ui.DecimalFragment;

public class MainActivity extends AppCompatActivity {

    private SectionsPagerAdapter mSectionsPagerAdapter;
    private ViewPager mViewPager;
```

```java
    @Override
    protected void onCreate(Bundle savedInstanceState) {
        super.onCreate(savedInstanceState);
        setContentView(R.layout.activity_main);

        Toolbar toolbar = (Toolbar) findViewById(R.id.toolbar);
        setSupportActionBar(toolbar);
        mSectionsPagerAdapter = new
SectionsPagerAdapter(getSupportFragmentManager());

        mViewPager = (ViewPager) findViewById(R.id.container);
        mViewPager.setAdapter(mSectionsPagerAdapter);

        FloatingActionButton fab = (FloatingActionButton) findViewById(R.id.fab);
        fab.setOnClickListener(new View.OnClickListener() {
            @Override
            public void onClick(View view) {
                ((CalculatorFragment)mSectionsPagerAdapter.getItem(mViewPager
.getCurrentItem()))).performCalculation();
            }
        });
    }

    public class SectionsPagerAdapter extends FragmentPagerAdapter {
        BinaryFragment bf=BinaryFragment.newInstance();
        DecimalFragment df=DecimalFragment.newInstance();
        public SectionsPagerAdapter(FragmentManager fm) {
            super(fm);
        }

        @Override
        public Fragment getItem(int position) {
            if (position==0)
                return  bf;
            else
                return df;
        }

        @Override
        public int getCount() {
            return 2;
        }

        @Override
        public CharSequence getPageTitle(int position) {
            switch (position) {
                case 0:
                    return "Binary";
                case 1:
                    return "Decimal";
            }
            return null;
        }
    }
}
```

LISTING 8-3: CalculatorFragment class code

```java
package com.expertandroid.chapter8.binary.ui;

public interface CalculatorFragment {
    void performCalculation();
}
```

LISTING 8-4: DecimalFragment class code

```java
package com.expertandroid.chapter8.binary.ui;

import android.os.Bundle;
import android.support.v4.app.Fragment;
import android.view.LayoutInflater;
import android.view.View;
import android.view.ViewGroup;
import android.widget.EditText;
import android.widget.TextView;
import com.expertandroid.chapter8.binary.R;
import com.expertandroid.chapter8.binary.logic.Calculator;

public class DecimalFragment extends Fragment implements CalculatorFragment{

    private EditText decimalNumber;
    private TextView binaryResult;
    private Calculator calculator;

    public DecimalFragment() {}

    public static DecimalFragment newInstance() {
        DecimalFragment fragment = new DecimalFragment();
        return fragment;
    }

    @Override
    public View onCreateView(LayoutInflater inflater, ViewGroup container,
                        Bundle savedInstanceState) {
        View v = inflater.inflate(R.layout.fragment_decimal, container, false);
        decimalNumber= (EditText) v.findViewById(R.id.decimalNumberEditText);
        binaryResult = (TextView) v.findViewById(R.id.binaryResultText);
        calculator=new Calculator();

        return v;
    }

    @Override
    public void performCalculation() {
        binaryResult.setText(calculator.convertToBinary(decimalNumber.getText()
.toString()));
    }
}
```

LISTING 8-5: BinaryFragment class code

```java
package com.expertandroid.chapter8.binary.ui;

import android.content.Context;
import android.os.Bundle;
import android.support.v4.app.Fragment;
import android.view.LayoutInflater;
import android.view.View;
import android.view.ViewGroup;
import android.widget.EditText;
import android.widget.TextView;
import com.expertandroid.chapter8.binary.R;
import com.expertandroid.chapter8.binary.logic.Calculator;

public class BinaryFragment extends Fragment implements CalculatorFragment{

    private EditText binaryNumber;
    private TextView decimalResult;
    private Calculator calculator;

    public BinaryFragment() {}

    public static BinaryFragment newInstance() {
        BinaryFragment fragment = new BinaryFragment();
        return fragment;
    }

    @Override
    public View onCreateView(LayoutInflater inflater, ViewGroup container,
                            Bundle savedInstanceState) {
        View v= inflater.inflate(R.layout.fragment_binary, container, false);
        binaryNumber= (EditText) v.findViewById(R.id.binaryNumberEditText);
        decimalResult = (TextView) v.findViewById(R.id.decimalResultText);
        calculator=new Calculator();

        return v;
    }

    @Override
    public void performCalculation() {
        decimalResult.setText(calculator.convertToDecimal(binaryNumber.getText()
.toString()));
    }

}
```

LISTING 8-6: Calculator class code

```java
package com.expertandroid.chapter8.binary.logic;

public class Calculator {
```

```
public String convertToBinary (String decimal){
    try{
        return Long.toBinaryString(Integer.parseInt(decimal));
    }catch (Exception e){
        return "Invalid input";
    }
}

public String convertToDecimal(String binary){
    try {
        return String.valueOf(Integer.parseInt(binary, 2));
    }catch (Exception e){
        return "Invalid input";
    }
}
```

Unit Tests

Unit tests are great for testing the functionality of a method. Good unit tests make bugs visible as they appear. The idea behind a unit test is to test a method with possible inputs and inspect the output. They are useful not only for typical cases. Unit tests for edge cases and error cases should also be written to provide good test coverage.

The Calculator class has two logic methods that are great candidates for unit testing. The project currently has two test folders: tests and androidTests. Right-click tests and create a new class as shown in Figure 8-49.

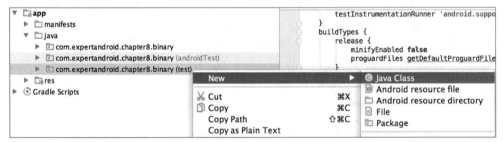

FIGURE 8-49: Creating a new class in the tests folder

Name the class `CalculatorTest`. Because you have two methods—one to convert binary to decimal and another to convert decimal to binary—you will write success and fail cases for both. For the sake of simplicity, you will write two tests for each method shown in Listing 8-7. (In real life, you may need to implement more test cases to cover edge cases, null input, extra large or small data sets, and other error conditions.)

LISTING 8-7: CalculatorTest class code

```
package com.expertandroid.chapter8.binary;

import com.expertandroid.chapter8.binary.logic.Calculator;
import org.junit.Before;
import org.junit.Test;
import static org.hamcrest.CoreMatchers.is;
import static org.junit.Assert.*;

public class CalculatorTest {

    private Calculator calculator;

    @Before
    public void initializeCalculator(){
        calculator=new Calculator();
    }

    @Test
    public void convertToBinarySuccess() throws Exception {
        assertThat(calculator.convertToBinary("256"), is("100000000"));
    }

    @Test
    public void convertToBinaryFail() throws Exception {
        assertThat(calculator.convertToBinary("12ww11"), is("Invalid input"));
    }

    @Test
    public void convertToDecimaSuccess() throws Exception {
        assertThat(calculator.convertToDecimal("1111"), is("15"));
    }

    @Test
    public void convertToDecimaFail() throws Exception {
        assertThat(calculator.convertToDecimal("121"), is("Invalid input"));
    }
}
```

> **NOTE** *Test-driven development works best when the test code is written without solid knowledge of how the method under test is implemented. Otherwise, the test may be biased and written in a way that makes it succeed. A great approach is writing the tests before writing the implementation because the input and output are already determined.*

Now you can run your test and see whether it succeeds. Right-click on the `CalculatorTest` class and select Run. Android Studio will run all tests in the class and display a summary of the status. Whether you are confident about your code or not, seeing the green bar shown in Figure 8-50 always feels great.

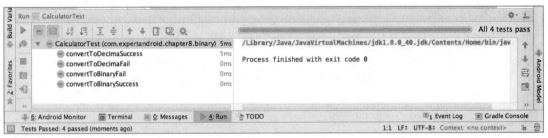

FIGURE 8-50: CalculatorTest results

This example used `assertThat` with `is` to test the output with expected values. There are several other methods that work for the same purpose:

➤ `assert`

➤ `assertEquals`

➤ `assertNotEquals`

➤ `assertTrue`

➤ `assertSame`

➤ `assertNotSame`

➤ `assertNull`

➤ `assertNotNull`

The use of these methods is beyond the scope of this chapter, but most `assert` methods expect two inputs: the output of the method under test and the expected value.

Any code that does not have UI interaction can be easily tested with basic unit tests. A good separation of UI and logic code will enhance testability, and several approaches such as MVC or MVP help you achieve this goal. Once the UI code is separated, the remaining code can be tested with simple unit tests.

Integration Tests

The method under test might be using an Android-specific API that needs a full-blown OS for running, for example, database operations on Android's SQLite. Starting with version 1.1 of the Android Gradle plugin, a mockable version of `android.jar` has been introduced in order to simplify testing. This approach can be used with popular mocking frameworks such as *mockito*. Our example will not be making use of powerful mockito features; however, in real life mockito greatly helps with mocking.

Integration tests do not need to open activities or fragments but can still test platform-specific APIs and features with the help of `android.jar`. They can also mock other integration points such as backend, database, or any other external resource. To run tests on a mockable version of `android.jar`, you would need to introduce product flavors, which we covered in Chapter 4.

Let's change our calculator to implement a sophisticated user history on a remote server based on social login. For the sake of simplicity, let's leave the implementation empty, as presented in Listing 8-8.

LISTING 8-8: History class code

```
package com.expertandroid.chapter8.binary.logic;

public class History {

    public History() {
        //Perform some fancy sophisticated social login
    }

    public void add(String item) {
        //add to remote cache
    }

    public String get() {
        //get from remote cache
        return null;
    }
}
```

Because our fancy user cache is ready, let's integrate it with the Calculator, as shown in Listing 8-9.

LISTING 8-9: Cache integration with Calculator

```
package com.expertandroid.chapter8.binary.logic;

public class Calculator {

    History history=new History();

    public String convertToBinary (String decimal){
        history.add(decimal);
        try{
            return Long.toBinaryString(Integer.parseInt(decimal));
        }catch (Exception e){
            return "Invalid input";
        }
    }

    public String convertToDecimal (String binary){
        history.add(binary);
        try {
```

```
            return String.valueOf(Integer.parseInt(binary, 2));
        }catch (Exception e){
            return "Invalid input";
        }
    }
}
```

Now, let's assume you don't want to go to the backend and make network calls simply to test the Calculator class. To achieve this goal, you need to mock the History class. Open the build .gradle file of the app module and add a product flavor for the purpose of mocking, as shown in Listing 8-10.

LISTING 8-10: Product flavors for mock

```
productFlavors {
    mock {
        applicationIdSuffix = ".mock"
    }
    prod {

    }
}
```

Now it is time to create your folder structure for the mock flavor. Navigate to the src folder and create a mock/java folder. Next, create a package with the same name as the package in which your History class resides: com.expertandroid.chapter8.binary.logic. Create a class named History and add the code shown in Listing 8-11.

LISTING 8-11: History mock implementation

```
package com.expertandroid.chapter8.binary;

import com.expertandroid.chapter8.binary.logic.History;

import java.util.Stack;

public class HistoryImpl implements History{
    private Stack<String> cache=new Stack<>();

    public void add(String item){
        cache.push(item);
    }

    public String get(){
        cache.pop();
        return "";
    }
}
```

Unlike the original `History` class, the mock `History` class uses an internal stack to keep values in memory instead of making network calls. Finally, in order to replace the original `History` class with the new mock `History` class, you need to move it to its own flavor. You already added prod behavior but you need to move the `History` class into a flavor-related directory structure. Navigate to the src folder and create a `prod/java` folder. Next, create a package with the name `com.expertandroid.chapter8.binary.logic` and move the original `History` class from the `main/java` folder to your newly created package. Now you can select either the prod or mock build variant to change the `History` class implementation in use, as shown in Figure 8-51.

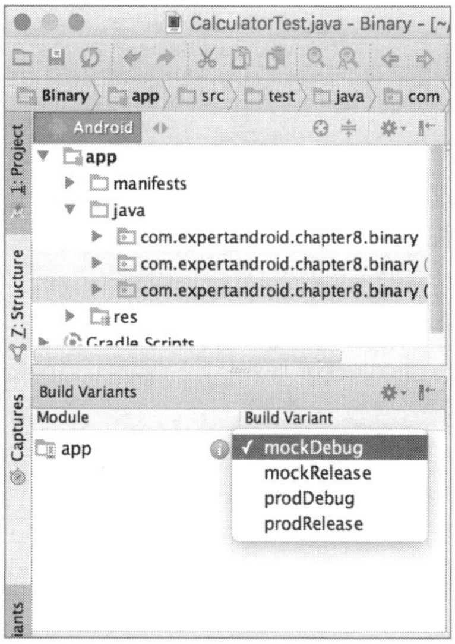

UI Tests

Testing the UI is another important step in the software development lifecycle. Even projects with pretty good unit test coverage fail to implement automated UI tests most of the time. UI tests provide complete end-to-end testing and help you figure out if your app is behaving as expected. Good UI test coverage would greatly help expose broken code or functionality. If you are coming from a web background or build HTML5-based apps for Android, you might already be familiar with automated web UI testing frameworks such WebDriver and Selenium.

FIGURE 8-51: Debugging a mock class

Android also has a powerful UI testing framework called "Espresso." The Espresso UI runs on an emulator or a device. Espresso can automate user interaction, fill fields, submit actions, and even analyze outputs and changes to make sure they are as expected. The final example of this chapter will test your UI by using your binary calculator, checking the output, and then switching to the decimal calculator and again running a calculation and testing the output.

Let's start by adding the dependencies needed for Espresso. Open your build.gradle file and add the dependencies shown in Listing 8-12.

LISTING 8-12: Espresso dependencies definition

```
apply plugin: 'com.android.application'

android {
    compileSdkVersion 23
    buildToolsVersion "23.0.2"

    defaultConfig {
        applicationId "com.expertandroid.chapter8.binary"
        minSdkVersion 23
        targetSdkVersion 23
```

```
            versionCode 1
            versionName "1.0"
            testInstrumentationRunner 'android.support.test.runner.AndroidJUnitRunner'
        }
    buildTypes {
        release {
            minifyEnabled false
            proguardFiles getDefaultProguardFile('proguard-android.txt'),
'proguard-rules.pro'
        }
    }
    productFlavors {
        mock {
            applicationIdSuffix = ".mock"
        }
        prod {
        }
    }
}

dependencies {
    compile fileTree(dir: 'libs', include: ['*.jar'])
    compile 'com.android.support:design:23.1.1'
    compile "com.android.support.test.espresso:espresso-idling-resource:2.2.1"
    testCompile "org.hamcrest:hamcrest-all:1.3"
    testCompile 'junit:junit:4.12'
    androidTestCompile 'com.android.support:support-annotations:23.1.1'
    androidTestCompile "com.android.support.test:runner:0.4.1"
    androidTestCompile "com.android.support.test:rules:0.4.1"
    androidTestCompile('com.android.support.test.espresso:espresso-core:2.2.1')
}
```

The parts highlighted in bold are newly added configurations and dependencies needed for the UI tests you are about to write.

Because you finished adding your dependencies, it is time to code. Let's start by listing the steps in the expected scenario and later turn them into code.

When the user launches your application, the first fragment in focus is the Binary calculator. The first step will be moving focus to the text field and writing a valid binary number. Next, you need to access the button and trigger an onClick event. Finally, you need to check that the displayed value is your expected result.

Now that you've decided on your testing scenario, go to the androidTest/java folder and create the com.expertandroid.chapter8.binary package if it does not already exist. Next, create a new class named ApplicationTest. Now you are ready to code. Listing 8-13 shows the first test, which tests the initial fragment.

LISTING 8-13: ApplicationTest code

```
package com.expertandroid.chapter8.binary;

import android.support.test.espresso.Espresso;
import android.support.test.rule.ActivityTestRule;
import android.support.test.runner.AndroidJUnit4;
import android.test.suitebuilder.annotation.LargeTest;

import org.junit.Rule;
import org.junit.Test;
import org.junit.runner.RunWith;

import static org.hamcrest.Matchers.allOf;
import static android.support.test.espresso.Espresso.onView;
import static android.support.test.espresso.action.ViewActions.click;
import static android.support.test.espresso.action.ViewActions.typeText;
import static android.support.test.espresso.assertion.ViewAssertions.matches;
import static android.support.test.espresso.matcher.ViewMatchers.isDisplayed;
import static android.support.test.espresso.matcher.ViewMatchers.withId;
import static android.support.test.espresso.matcher.ViewMatchers.withText;

@RunWith(AndroidJUnit4.class)
@LargeTest
public class ApplicationTest {


    @Rule
    public ActivityTestRule<MainActivity> mActivity =
            new ActivityTestRule<>(MainActivity.class);

    @Test
    public void calculateBinary() throws Exception {
        // type the number
        onView(allOf(withId(R.id.binaryNumberEditText),
isDisplayed())).perform(typeText("111"));

        Espresso.closeSoftKeyboard();//
        // perform the click
        onView(withId(R.id.fab)).perform(click());

        // check the output
        onView(withText("7")).check(matches(isDisplayed()));
    }
}
```

The first part of the class file is the `ActivityTestRule` declaration. A rule provides functional testing of a single activity by launching the activity for each method annotated with `@Test`.

Next you have your test method `calculateBinary`, which is annotated with `@Test`. This method consists of four lines. The first line looks for the field `binaryNumberEditText` with the field ID on a view. However, because your fragments are in a view pager, you would need to look for the one you want by using `isDisplayed` matcher. Although you field ID is unique, it is in a adapter so there are no guarantees that there are no other similar instances. Once you find the `binaryNumberEditText` field, you perform a type text action to enter the value to test. Now you need to click the button, but as in real life, the software keyboard is currently covering the button. The `Espresso` `.closeSoftKeyboard` method closes the keyboard. Next, you can click the button by using its ID and finally search for the expected value in the displayed field.

You have finished your first UI test. Now either launch the emulator or connect a device and run the test by right-clicking and selecting the Run ApplicationTest option as shown in Figure 8-52.

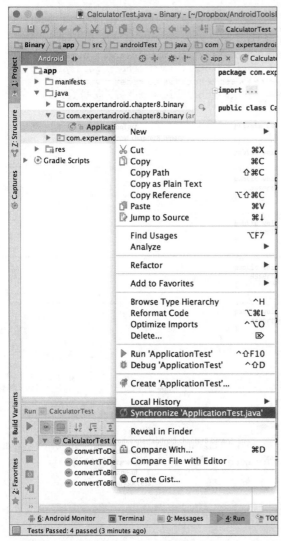

FIGURE 8-52: Running an Espresso test

Now that you have finished testing the binary fragment, you can move on to testing the decimal fragment. `DecimalFragment` is almost an identical copy of `BinaryFragment`. So you will use similar code but change UI field names and values to test the `DecimalFragment`. However, you also need to swipe left and display the `DecimalFragment` first, as in Listing 8-14.

LISTING 8-14: Swipe test addition to ApplicationTest

```java
package com.expertandroid.chapter8.binary;

import android.support.test.espresso.Espresso;
import android.support.test.espresso.action.GeneralLocation;
import android.support.test.espresso.action.GeneralSwipeAction;
import android.support.test.espresso.action.Press;
import android.support.test.espresso.action.Swipe;
import android.support.test.rule.ActivityTestRule;
import android.support.test.runner.AndroidJUnit4;
import android.test.suitebuilder.annotation.LargeTest;

import org.junit.Rule;
import org.junit.Test;
import org.junit.runner.RunWith;

import static org.hamcrest.Matchers.allOf;
import static android.support.test.espresso.Espresso.onView;
import static android.support.test.espresso.action.ViewActions.click;
import static android.support.test.espresso.action.ViewActions.typeText;
import static android.support.test.espresso.assertion.ViewAssertions.matches;
import static android.support.test.espresso.matcher.ViewMatchers.isDisplayed;
import static android.support.test.espresso.matcher.ViewMatchers.withId;
import static android.support.test.espresso.matcher.ViewMatchers.withText;

@RunWith(AndroidJUnit4.class)
@LargeTest
public class ApplicationTest {


    @Rule
    public ActivityTestRule<MainActivity> mActivity =
            new ActivityTestRule<>(MainActivity.class);

    @Test
    public void calculateBinary() throws Exception {
        //...
    }

    @Test
    public void calculateDecimal() throws Exception {
        onView(withId(R.id.container)).perform(new
GeneralSwipeAction(Swipe.FAST, GeneralLocation.CENTER_RIGHT,
                GeneralLocation.CENTER_LEFT, Press.FINGER));

        // type the number
        onView(allOf(withId(R.id.decimalNumberEditText),
```

```
isDisplayed())).perform(typeText("7"));

    Espresso.closeSoftKeyboard();//
    // perform the click
    onView(withId(R.id.fab)).perform(click());

    // check the output
    onView(withText("111")).check(matches(isDisplayed()));
  }
}
```

The first line in the test performs the swipe action by creating a new `GeneralSwipeAction` object. `GeneralSwipeAction` has the following parameters in order: swipe type, start location, end location, and input device that triggered the swipe.

Now you have all the pieces you need so you just need to tie them together in order to run them as a whole large test. JUnit 4 has great utilities to accomplish your goal. Check that you have the annotations in the following code snippet, above your class declaration.

```
@RunWith(AndroidJUnit4.class)
@LargeTest
```

Run the application test and watch how the application ran and what interaction occurred on the device or emulator. Finally, both tests should pass and display something similar to Figure 8-53.

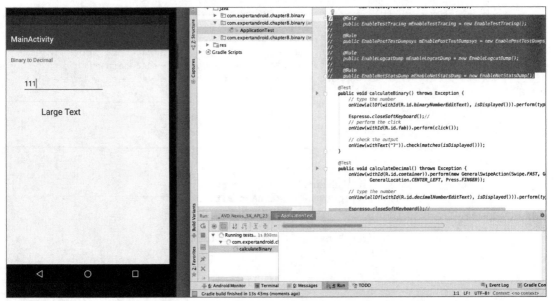

FIGURE 8-53: Test output

That's it—you wrote unit and UI tests and integrated them with your existing code. The next section focuses on testing your app's performance and collecting valuable information and statistics.

Performance Testing

So far, you tested the functionality of your methods and user interaction with the UI, but there is another important aspect of testing. Performance tests can reveal memory leaks and unnecessary computation or object initialization, which result in performance problems in applications. Such problems do not appear in unit, integration, or UI tests.

The first tool you are going to use for testing performance is Systrace, which captures and displays the application code execution time. To start Systrace, click on the terminal tab in Android Studio and type the following command, depending on your OS. Listing 8-15 shows the command for Windows, and Listing 8-16 is for Linux and Unix.

LISTING 8-15: Systrace terminal command on Windows

```
$ python %ANDROID_HOME%/platform-tools/systrace/systrace.py --time=XX -o
%userprofile%/trace.html gfx view res
```

LISTING 8-16: Systrace terminal command on Linux and Unix

```
$ python $ANDROID_HOME/platform-tools/systrace/systrace.py --time=XX -o
~/trace.html gfx view res
```

Replace *xx* with the number of seconds you want your app to be traced. Now you can open a browser and view the `trace.html` document produced by the Systrace (see Figure 8-54).

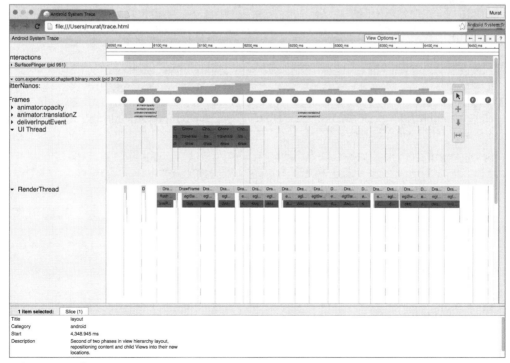

FIGURE 8-54: Systrace output

The Systrace report is interactive and enables users to focus on items to examine details, zoom in or out, or highlight the timeline. Click on the alerts line to display alerts raised by the app. Most of the time, alerts can provide very useful clues about performance bottlenecks.

Performance Tests Task

Running Systrace to examine your app performance is an essential method to understand performance problems. Watching the performance changes by automating Systrace is crucial. You have already covered automated UI testing in previous sections; now you will add Systrace to accompany them.

Google has revealed a set of new test rules with an android-perf-testing codelab. These rules might become part of a new test support library or dependency in the near future. Until then, it is okay to check out and copy the classes from `https://github.com/googlecodelabs/android-perf-testing/tree/master/app/src/androidTest/java/com/google/android/perftesting/testrules` to your project. These rules can easily be added to your test projects, as shown in Listing 8-17.

LISTING 8-17: Performance test rules

```
    @Rule
    public EnableTestTracing mEnableTestTracing = new EnableTestTracing();

    @Rule
    public EnablePostTestDumpsys mEnablePostTestDumpsys = new
EnablePostTestDumpsys();

    @Rule
    public EnableLogcatDump mEnableLogcatDump = new EnableLogcatDump();

    @Rule
    public EnableNetStatsDump mEnableNetStatsDump = new EnableNetStatsDump();
```

Add the rules in Listing 8-17 to `ApplicationTest`. You should also add the `@PerfTest` annotation to the class to enable it for performance tests. Finally, open the build.gradle file and add the following line to apply performance tasks:

```
    apply plugin: PerfTestTaskGeneratorPlugin
```

Your final code should look like Listing 8-18.

LISTING 8-18: Performance test addition to ApplicationTest

```
    package com.expertandroid.chapter8.binary;

    import android.support.test.espresso.Espresso;
    import android.support.test.espresso.action.GeneralLocation;
    import android.support.test.espresso.action.GeneralSwipeAction;
    import android.support.test.espresso.action.Press;
    import android.support.test.espresso.action.Swipe;
```

```java
import android.support.test.rule.ActivityTestRule;
import android.support.test.runner.AndroidJUnit4;
import android.test.suitebuilder.annotation.LargeTest;

import org.junit.Rule;
import org.junit.Test;
import org.junit.runner.RunWith;

import static org.hamcrest.Matchers.allOf;
import static android.support.test.espresso.Espresso.onView;
import static android.support.test.espresso.action.ViewActions.click;
import static android.support.test.espresso.action.ViewActions.typeText;
import static android.support.test.espresso.assertion.ViewAssertions.matches;
import static android.support.test.espresso.matcher.ViewMatchers.isDisplayed;
import static android.support.test.espresso.matcher.ViewMatchers.withId;
import static android.support.test.espresso.matcher.ViewMatchers.withText;

@RunWith(AndroidJUnit4.class)
@LargeTest
@PerfTest
public class ApplicationTest {

    @Rule
    public EnableTestTracing mEnableTestTracing = new EnableTestTracing();

    @Rule
    public EnablePostTestDumpsys mEnablePostTestDumpsys = new
EnablePostTestDumpsys();

    @Rule
    public EnableLogcatDump mEnableLogcatDump = new EnableLogcatDump();

    @Rule
    public EnableNetStatsDump mEnableNetStatsDump = new EnableNetStatsDump();


    @Rule
    public ActivityTestRule<MainActivity> mActivity =
            new ActivityTestRule<>(MainActivity.class);

    @Test
    public void calculateBinary() throws Exception {
        //...
    }

    @Test
    public void calculateDecimal() throws Exception {
        onView(withId(R.id.container)).perform(new
GeneralSwipeAction(Swipe.FAST, GeneralLocation.CENTER_RIGHT,
                GeneralLocation.CENTER_LEFT, Press.FINGER));

        // type the number
        onView(allOf(withId(R.id.decimalNumberEditText),
isDisplayed())).perform(typeText("7"));
```

```
        Espresso.closeSoftKeyboard();//
        // perform the click
        onView(withId(R.id.fab)).perform(click());

        // check the output
        onView(withText("111")).check(matches(isDisplayed()));
    }
}
```

When executed, your test will collect and log more information to help you fix performance problems.

SUMMARY

This chapter gave you a solid, basic understanding of the full power of Android Studio's testing capabilities and debugging tools so you can debug and test your Android code efficiently.

We started by discussing debugging and the underlying technology, adb, which allows you to remotely debug Android devices. We followed up with Android Monitor in Android Studio, which is used for better monitoring and system information capturing.

Next, we looked at the Android Device Monitor for further Android device and application monitoring, and at legacy tools.

We finished the chapter working on test methods and tools for Android applications using Android Studio and SDK tools. You learned how to use automated tests of your application's functionality and UI as well as how to use performance tests to collect and record valuable runtime information.

Using Source Control: GIT

WHAT'S IN THIS CHAPTER?

- ➤ Sharing your project to source control
- ➤ Using Git on the command line
- ➤ Third-party tools for Git
- ➤ Android Studio Git integration

Whether you are using Git or something else, the source control system is an important part of the software development lifecycle. Whether you are working in a team or on your own, source control systems provide a full history of what you have been doing and ease the management of changes to and versioning of the code.

Popular source control systems, both free or paid and open source or vendor-based, have been around for quite a while. CVS was one of the early popular source control systems that many developers have been familiar with since the early 2000s. Later, SVN became a popular and widely used source control system, mostly replacing CVS thanks to its transactional commits. Commercial source control systems also emerged, such as IBM's ClearCase and Microsoft's SourceSafe.

However, the search for better source control did not end. Mercurial and Git became available around the same time, both addressing the same issue: distributed source control. From its introduction, Git has been widely accepted by many large open source projects, such as Linux and Eclipse. With the help of GitHub, which hosts many open source projects, Git has been the de facto standard for source control.

This book covers Git because it is currently the most widely accepted source control system among Android developers, and it has built-in support for Android Studio.

INTRODUCTION TO GIT

The idea behind source control is simple. Source control systems save changes as patches (commits). This provides an easy way to revert to or compare points in the development timeline. This history of changes not only provides insight based on what has happened in this project but also provides a great way to integrate different development efforts.

Changes are performed locally and committed to a central server, which acts as the version control server. This architecture does not allow developers to work offline and relies on a persistent connection between the local source files and the version control system. However, Git is different; it is distributed. Unlike other systems that just watch changes in the file system and commit changes to the version control server, Git runs on a client computer and changes need to be committed locally first. This way, a developer can revert any change/branch or version locally through Git. The local Git can push the set of changes committed to network Git servers. All changes are kept both locally and on any number of servers, as shown in Figure 9-1, which makes Git very flexible and powerful. You can create any combination of branches for teams.

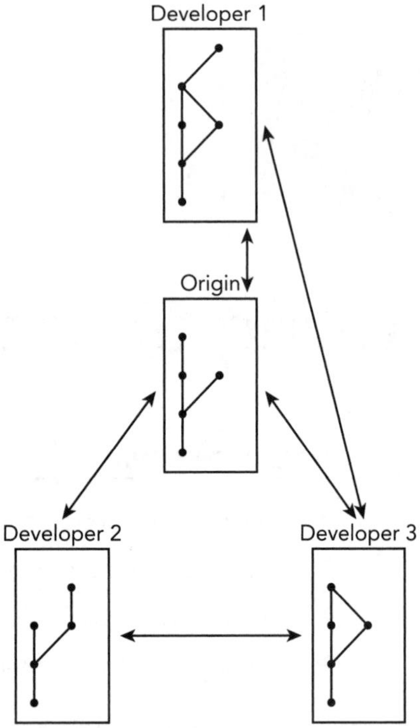

FIGURE 9-1: Distributed version control using Git

Understanding Git

The basic idea of source control is to keep a stable version of the code, usually called the *master* branch. Ideally, any development effort should be performed in a separate branch so you don't break the master. Changes are committed in patches to the branch as they are completed. The commit patches include a bunch of changes, which are wrapped in an atomic transaction. Therefore, a change is either accepted or rolled back as a whole. Once the development purpose of the branch is complete and tested, it is merged with the master. The process is shown in Figure 9-2.

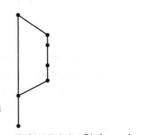

FIGURE 9-2: Git branch

Ideally, a branch, consisting of a batch of commits, is merged to the master smoothly by a triggered pull request. However, in the real world, that is not the case most of the time. Some other development branch might have been merged before your branch or a critical bug might have been patched and merged into the master. Because the entire idea behind version control is to enable teams to work in harmony, this should not be an issue.

Figure 9-3 shows that branch B has been merged before branch A. Ideally, this merge happens smoothly without any conflict, provided that the two development branches were different files or different parts of the same file without creating a conflict. However, once again this is not often the case in the real world. Other developers may need to patch a critical bug, or they may be developing a feature that overlaps with the code fragment you are working on, which creates a conflict. Figure 9-4 shows a conflict scenario.

A conflict happens when there's a change in a line or the same segment of code in two or more parallel but different timelines. When it is time to combine these changes, manual action is required because, although both were meaningful when the branchout occurred, one might become obsolete or create a need for additional change in the other. To resolve this conflict, one or more developers who have an understanding of both braches need to compare both versions of the code to perform a manual merge of those and then merge the merged code into the master. Of course, this is a simplified description of this operation and there are other process flows you may choose to follow.

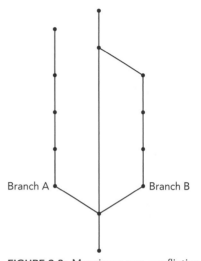

FIGURE 9-3: Merging a non-conflicting branch

You can use Git either while starting the project or at a later stage in development. This chapter covers both options, but sharing your project to Git from the start is definitely the preferred way.

Installing Git

Android Studio comes with Git support. However, you may still need to install Git to be able to use it through the command line. You can install Git as follows, depending on your OS.

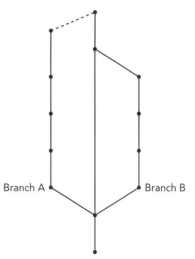

FIGURE 9-4: Merging conflicting branch

➤ **MacOsX**—A binary Git installer for Mac OS X is available at `http://git-scm.com/download/mac`. An alternative way to install Git is to install GitHub Client for Mac OS X, which also offers a simplified user interface for Git at `http://mac.github.com/` (see Figure 9-5).

➤ **Linux**—Popular package managers all support binary Git installation. To install Git on Debian-based systems (such as Ubuntu), use the following command:

```
sudo apt-get install git-all
```

Alternatively, you can install Git via yum with this command:

```
sudo yum install git-all
```

➤ **Windows**—Just like Mac OS, there is a binary installer for Windows. It's available at `http://git-scm.com/download/win`. You may prefer to install GitHub client for

Windows, which comes with a simplified user interface. It's available at `http://windows` `.github.com/`.

Once the installation is complete, you need to create your identity. To set your name and email address, execute the following commands directly from the terminal.

```
$ git config --global user.name "John Doe"
$ git config --global user.email johndoe@example.com
```

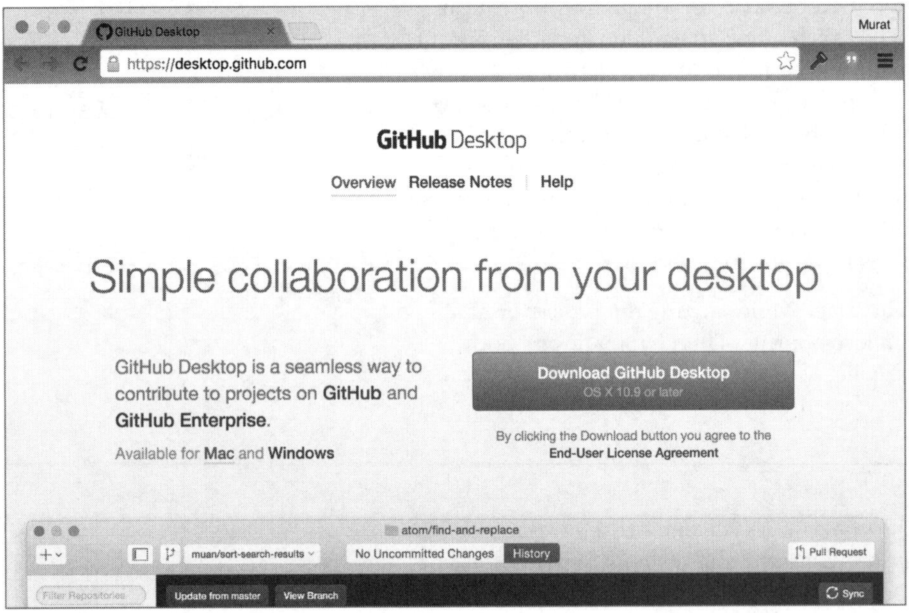

FIGURE 9-5: GitHub page

> **TIP** *On both Mac OS X and Linux, you can just run the Terminal application to access the shell and run the preceding commands. If you are using Microsoft Windows, you should run the Git Bash app, which is installed together with Git.*

That's it: You installed Git and are now ready to create a project.

USING GIT

Let's start from scratch by sharing a project to Git via the command line:

1. Navigate to your target project's folder from the terminal with the following command.

```
$ cd /path/to/projectfolder
```

2. Create a new folder to hold the source files and share them to Git.

```
$ mkdir git-project
```

3. Next, you need to add this folder to Git. Start by initializing Git:

```
$ git init
```

> **NOTE** *If you get an error message, as shown in Figure 9-6, please refer to the previous section on how to set your Git identity.*

```
Your name and email address were configured automatically based
on your username and hostname. Please check that they are accurate.
You can suppress this message by setting them explicitly:

    git config --global user.name "Your Name"
    git config --global user.email you@example.com

After doing this, you may fix the identity used for this commit with:

    git commit --amend --reset-author

1 file changed, 1 insertion(+)
create mode 100644 read.me
```

FIGURE 9-6: Git configuration output

That's it! This folder is registered as a Git root. Now you can add folders and files. However, do not forget that Git does not just track the file system. Adding files to this directory does not mean anything unless the files are also added to Git. Create a read.me file for your git-project folder and run the following command to add that file to Git.

```
$ git add read.me
```

The file is added to Git and is being tracked. However, to start versioning, you need to commit changes to Git. The addition of the file is an *initial* commit.

```
$ git commit -m 'initial commit'
```

That's it! Edit the file, make some changes, and perform another commit. Use the following command to check the file's status. This command will list uncommitted changes if there are any.

```
$ git status
```

But wait a minute! You are still performing all the versioning on your local machine, so another developer cannot access your source file. Worse, what if something happens to your computer? To push the changes to a remote Git repository, you first need to find a Git host. GitHub is a popular Git service, and you will use it in this chapter. GitHub offers free unlimited repositories for open source projects and it has reasonable pricing options if you want private repositories.

If you do not have a GitHub account, visit http://www.github.com and create an account before proceeding to the next section.

USING THE GITHUB CLIENT

GitHub is a popular Git-based project-hosting site that offers free hosting for public repositories. One reason for GitHub's popularity is the available easy-to-use tools for Git. The tools proved to be so successful at being simple and easy to use that some people even organized their weddings with GitHub.

GitHub provides tools for all major operating systems. To get installers or the application binary, navigate to `https://desktop.github.com/`. When you are there, it will recognize your operating system. If you are using Mac OS X, it will download a zip file named `GitHub Desktop 220.zip` (220 is the version number).

Mac OS X installation is easy; just extract the zip file and copy `GitHub Desktop.app` to your application's directory and installation is done.

If you are using Windows, the website recognizes that and will download the `GitHubSetup.exe` file. To start installation, launch that file.

When the security notification shown in Figure 9-7 appears, click Install.

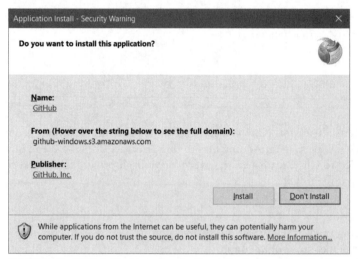

FIGURE 9-7: GitHub Desktop Application Install – Security Warning window

As shown in Figure 9-8, the installer downloads the GitHub Desktop application files.

FIGURE 9-8: GitHub Desktop installation process

When the installation finishes, the Welcome window shown in Figure 9-9 comes up. (This figure shows the Windows version of this screen.)

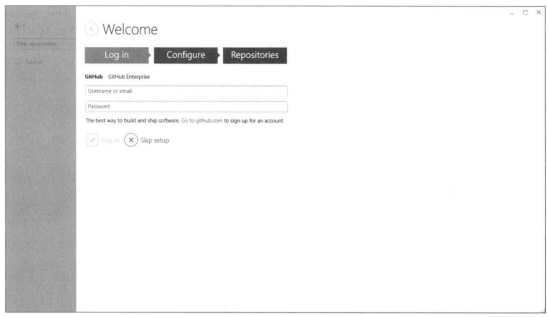

FIGURE 9-9: GitHub Desktop start screen

After installing GitHub Desktop on Mac OS X, start the application to see a window with the same functionality as in Windows. Figure 9-10 shows this screen.

FIGURE 9-10: GitHub Desktop welcome screen

Clicking Skip Setup takes you to the main screen. If you click Continue, you will need to enter your GitHub username and password, as shown in Figure 9-11. GitHub Desktop can work without GitHub and even with other Git hosting sites. However, completing your GitHub account setup will bring up all the GitHub repositories you previously created with that account.

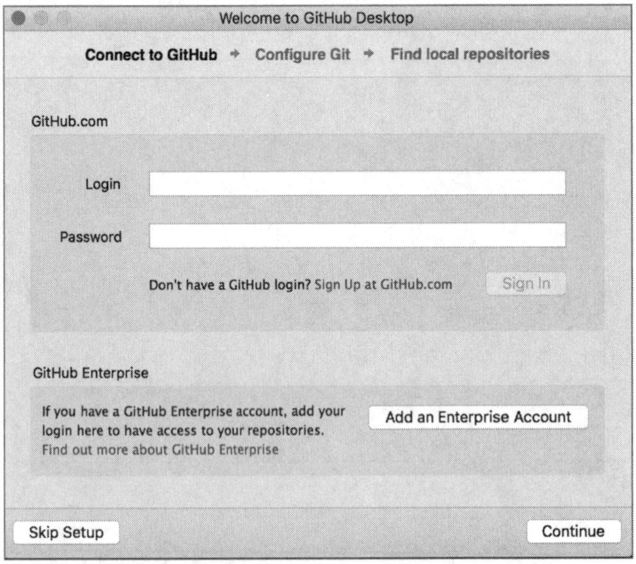

FIGURE 9-11: Connecting to GitHub

A typical setup with no repositories will look like Figure 9-12. This screen will differ depending on the specific local repositories or GitHub repositories added with your setup.

Next, let's focus on how to add existing repositories or create new ones using GitHub Desktop. Let's start with creating a new repository. Select New Repository from the File menu or click the plus (+) sign in the upper-left corner (see Figure 9-13).

You will be asked for the parent directory where you want your local repository, as shown in Figure 9-14. By default, this directory is the GitHub folder in your home directory. You may continue with the default selection or enter your desired folder's name.

> **NOTE** *The folder structure you use to organize your projects is a personal pref-erence. If you want to keep all your projects in one place, you can change the default folder to anything you prefer or even create subfolders based on your project types or where they are hosted.*

GitHub Desktop Client will create and initialize your folder and add it to the Other projects tab on the left. The project is listed under Other because it is not a GitHub project, as shown in Figure 9-15.

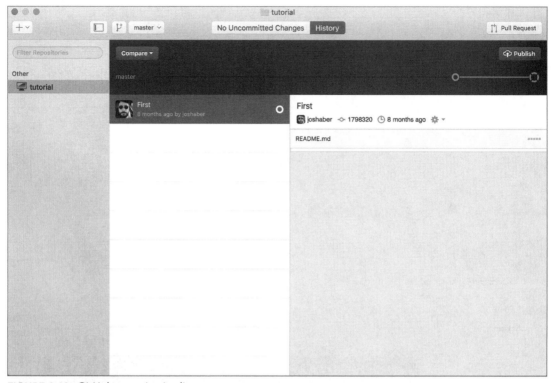

FIGURE 9-12: GitHub repositories list

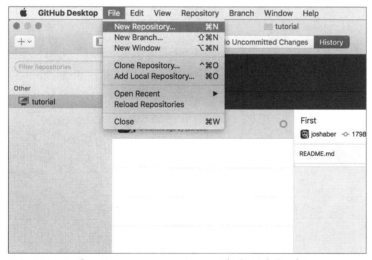

FIGURE 9-13: Creating a new repository with GitHub Desktop

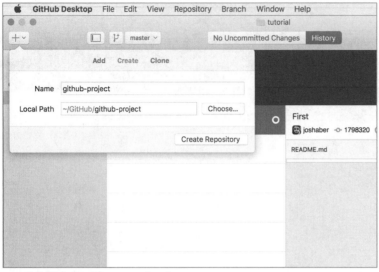

FIGURE 9-14: Repository path selection

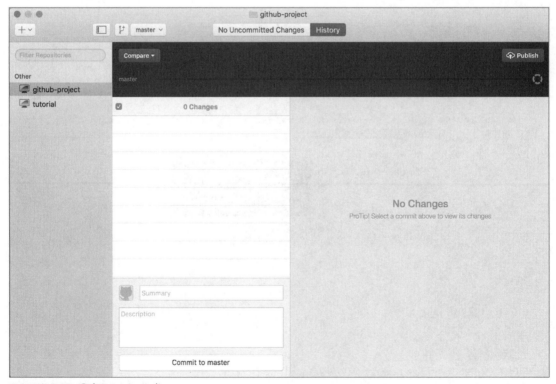

FIGURE 9-15: Other projects list

Now that you've initialized your project, it is time to make and track some changes. Copy the `read` `.me` file you previously placed in your git-project folder to your new Git repository. The GitHub Desktop Client will immediately show untracked changes, as shown in Figure 9-16.

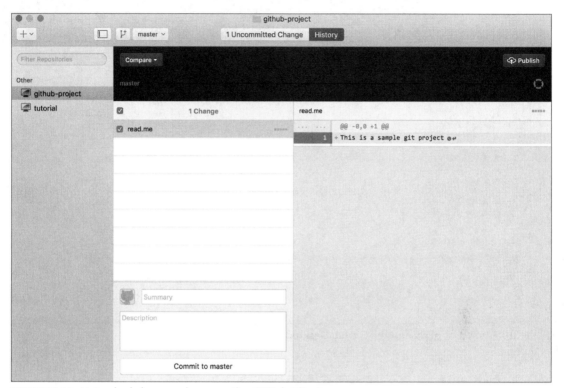

FIGURE 9-16: Untracked changes shown on the GitHub Client

The tracked and untracked changes to files are listed in the middle section of the github-project window. As shown in Figure 9-17, if you highlight a file, the change details are shown in the right pane. Lines that start with a plus (+) sign and are highlighted with green are newly added lines. Lines that start with the minus (–) sign and are highlighted with red are deleted lines. To commit this change, type a meaningful description and click the Commit to master button (refer to Figure 9-16). Since this is the initial commit, as the description, type **initial commit**.

You've created a new project and committed some changes through GitHub Client, but now let's work with your existing Git repositories via GitHub Desktop Client. This time, let's import your first git-project.

You can add any existing local or remote Git project. Click the + sign in the top-left corner and select Add, as shown in Figure 9-18.

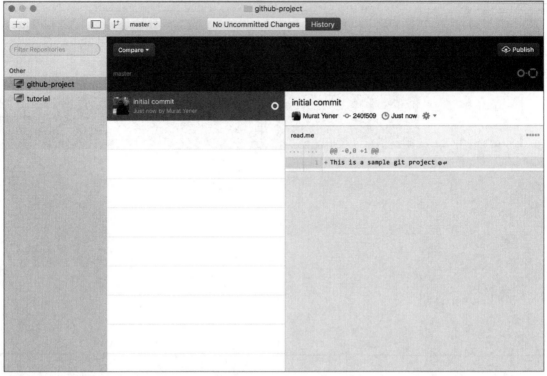

FIGURE 9-17: List of changes on GitHub Desktop

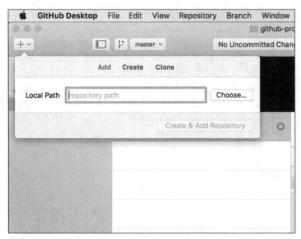

FIGURE 9-18: Importing a project

Next, type the path to your local folder or browse and navigate to the root folder of your Git project, as shown in Figure 9-19.

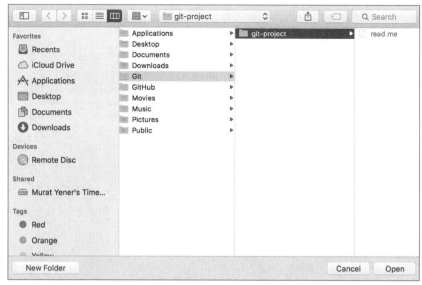

FIGURE 9-19: Git project path

Once the project is added, the whole history as well as all remote repository settings will be shown inside the GitHub client. You can continue working from the GitHub client or from the command line, or even switch back and forth.

You have created Git projects from both the command line and the GitHub client, and then made some changes and committed them, but no Git project is complete without a remote repository. By adding a remote repository, you enable other developers to work on the same project, and most importantly, you ensure the project is not saved on a local hard drive. To add a remote repository on GitHub, click the Publish button at the top-right corner. If you have not yet completed your GitHub sign-in, the warning shown in Figure 9-20 appears.

FIGURE 9-20: Sign-in warning

Click OK to open a new screen where you enter your GitHub username and password, as displayed in Figure 9-21.

FIGURE 9-21: Sign-in screen

When you have completed the sign-in process, you are asked for the name and description of a repository. After you've entered that information, click the Publish Repository button, as shown in Figure 9-22.

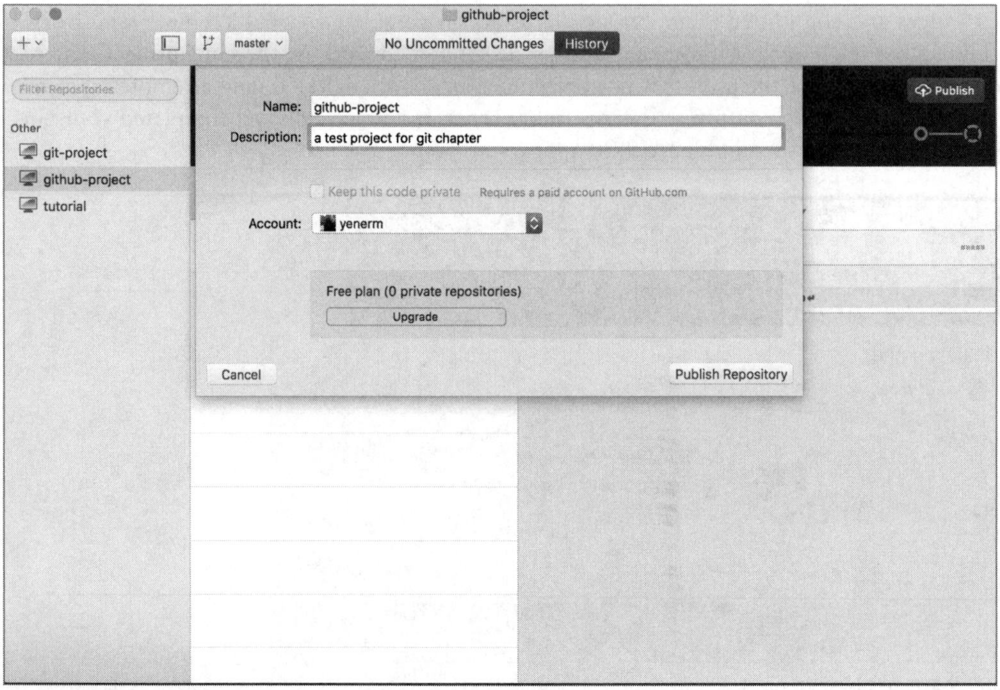

FIGURE 9-22: Entering the repository name and description

It's done! Your project is shared and hosted in a repository on GitHub. You can confirm that the project is now listed under GitHub projects in the left pane, as displayed in Figure 9-23.

FIGURE 9-23: GitHub project display

We have covered the difference between committing and pushing. The Sync button, shown in Figure 9-24, pushes local changes while pulling remote changes, if any, to your remote repository and keeps the changes synced.

You have seen using Git both from the command line and via the GitHub client to create and share projects. Now let's move to Android Studio and see how it is integrated with Git.

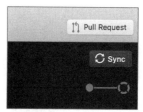

FIGURE 9-24: The Sync button on GitHub

USING GIT IN ANDROID STUDIO

This chapter has covered how to create a Git repository, add files to it, and perform commits. However, you need to know how to do much more than that for daily development tasks, including branching, merging, resolving conflicts, ignoring untracked files, and so on. This section covers Git using Android Studio, as well as switching to the command line and using third-party tools when needed.

You can always choose to use Git via the command line or a third-party tool, but having integrated support in your IDE may simplify your job. Indeed, Murat's favorite Git tool during his early days was e-git, which is actually just a Git integration plugin written for Eclipse. There is nothing wrong with using whichever tool works best for you.

In this section, you begin by moving an Android project to Git. To do this, you will create a new project, as shown in Figure 9-25, but you can use any Android project you have worked on as long as it hasn't been shared to Git before. The project you are going to follow has phone/tablet and wear modules with empty default activities.

You will be using the com.expertandroid.git package name. Also, feel free to choose different SDK versions or a different number of modules (see Figure 9-26) because this chapter focuses only on the Git flow of the project.

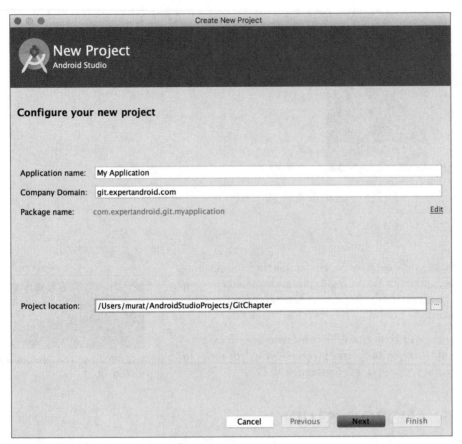

FIGURE 9-25: Creating a new project

Once the project is created, highlight it and select Enable Version Control Integration from the VCS (version control system) menu, as shown in Figure 9-27.

A popup dialog box will ask which version control system you want to integrate; choose Git. Android Studio has support for all major VCSs such as CVS, Subversion, and Mercurial, as you can see in the list in Figure 9-28.

Alternatively, you may use the options shown when you click the Import into Version Control item from the VCS menu (refer to Figure 9-27) to enable Git or another version tracking system (see Figure 9-29).

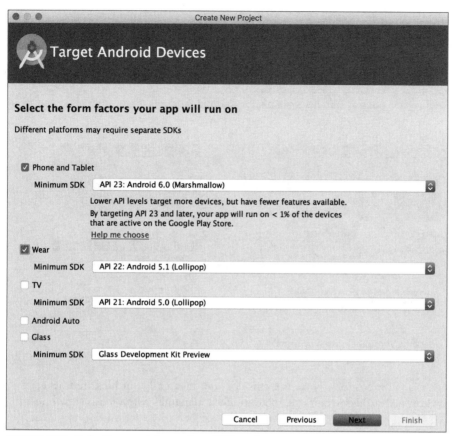

FIGURE 9-26: Selecting modules and an SDK version

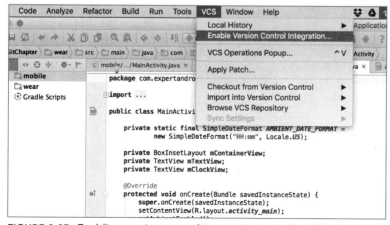

FIGURE 9-27: Enabling version control integration in Android Studio

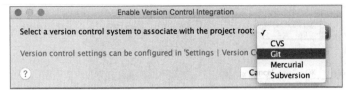

FIGURE 9-28: List of version control systems available for Android Studio

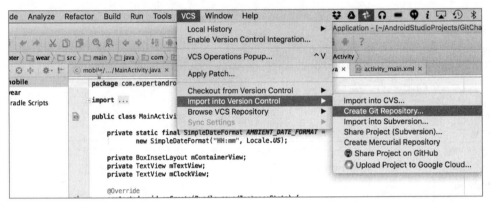

FIGURE 9-29: Import into Version Control options

Now your module is a Git repository. However, this does not mean all your files are tracked. Android Studio warns about this by marking the untracked filenames with red, as shown in Figure 9-30.

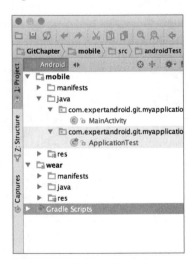

FIGURE 9-30: Untracked changes in Android Studio

As you remember from what you have done previously via the command line or the GitHub client, initializing a Git repository is only the first step. After the initialization, you need to add the files you want to track and then perform an initial commit. So now let's add your files to your repository. Select the module and right-click to display actions. Select Add from the Git option, as shown in Figure 9-31. Alternatively, you can press Shift+Command+A (Ctrl+Alt+A on Windows).

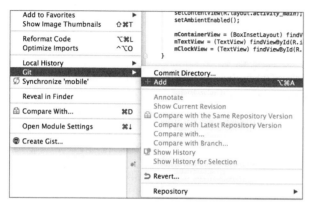

FIGURE 9-31: Adding items to the Git repository

This operation can be performed on files, packages, modules, or even the whole project. After they've been added (as shown in Figure 9-32), the filenames change from red to bright green in Android Studio's Project View.

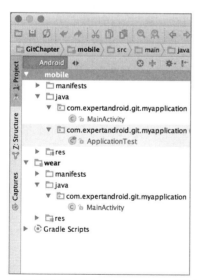

FIGURE 9-32: Added changes to Git in Android Project View

You are almost there: You've initialized your repository, added your files, and now it is time to commit your changes. Once again, highlight your module, then right-click and select Commit Directory from the Git option, as displayed in Figure 9-33.

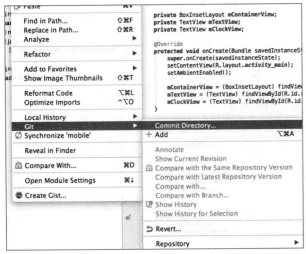

FIGURE 9-33: Commit action in Android Studio

This should bring up a commit wizard to help you choose multiple files and commit options as well as the commit message, as shown in Figure 9-34.

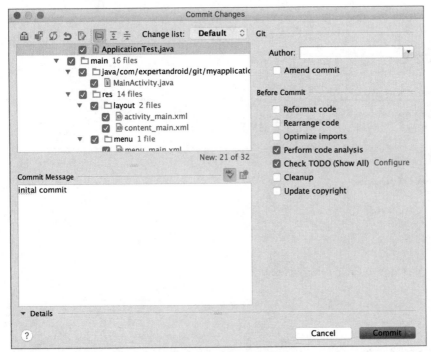

FIGURE 9-34: Android Studio Commit Changes wizard

This wizard window has three main components. The top-left window displays the file tree and enables you to choose multiple files to include in the commit. Do not worry about the icons on top for now; we will cover them in the following section. The bottom-left window has a multiline text entry field for the commit message. Composing a meaningful and self-explanatory commit message is not only a good practice but also necessary if you want your Git history to be browseable and meaningful. Because this is the initial commit, a simple initial commit message should be sufficient.

> **TIP** *In practice, you would need to write self-explanatory commit messages. If the commit includes development of a new feature, add the task number and includ a simple explanation of the added feature. Most Task Management Systems can be linked within Git commits. For example, to link a Jira task numbered 208 in a commit, you can type #208.*
>
> *Similarly, if the commit includes a fix for a bug, the tracking ID, explanation of the bug, and the resolution should be added to the commit message.*
>
> *It might be a good idea to include a simplified version of your task/bug explanation text into the commit if your task management system and Git repository do not support linking to each other.*

The right part of the commit wizard offers the following options:

➤ **Author**—The name of the author of the commit.

➤ **Amend Commit**—This option enables you to add changes to your previous commit (if, for example, you need to fix something in it). Unlike most other VCSs, this option allows you to change the contents of the previous transaction without rolling back the full commit.

➤ **Reformat Code**—Android Studio offers great code formatting options, which we covered in Chapter 4. By selecting this checkbox, you are asking Android Studio to reformat each applicable file before the commit.

➤ **Rearrange Code**—Works just like the Reformat Code option. Rearranges code before the commit.

➤ **Optimize Imports**—Removes unnecessary imports from the source file before the commit. While the unused imports are harmless and will be removed by the compiler, this option results in a shorter and relatively easy-to-read source file. On the other hand, this might be handy only when you read the source code in a simple text editor because Android Studio already hides imports.

➤ **Perform Code Analysis**—Analyzes the code and runs `lint` with given options. This option greatly helps to eliminate poorly written and problematic code, preventing it from being committed into the written history of the project.

➤ **Check TODO**—Just like the previous option, the Check TODO option helps prevent committing unfinished code into your repository. Although technically any code that is committed needs to be complete, that doesn't mean the code should always be TODO free. One may have added a TODO for a feature task, which would be the subject of another commit.

➤ **Cleanup**—Works pretty much like Reformat Code. This option runs cleanup rules before the commit.

➤ **Update Copyright**—Although it may seem not important, this is a very handy option, especially if you are working on a corporate or an open source project that relies heavily on properly placed copyright licenses. This option lets Android Studio update copyright licenses in each committed file.

Now that you've learned about all of the options, let's try a commit.

Clicking the Commit button reveals the options shown in Figure 9-35. You may choose to Commit, Commit and Push, or Create a Patch from changes. Click Commit to perform just a commit.

FIGURE 9-35: The Commit button in Android Studio

Because the generated code in the project introduces TODOs and warnings, your commit should pause with a warning, as shown in Figure 9-36.

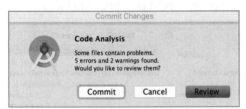

FIGURE 9-36: Code Analysis warning window

If you select Review, Android Studio will display all warnings, TODOs, and potential problems. Typically, it is good practice to click Review and go over all items to ensure they will not introduce any future problems. However, because the project currently contains only generated code, go ahead and click Commit. That is it; all source files in your project should have turned black from green when the commit is completed.

That's it; you have completed your `init`, change, and commit in your repository. This was simple, right? Well, of course, in real life you would face conflicts and merge problems. The next section discusses how to handle those.

GIT FLOW

During the development lifecycle, developers work on tasks that may or may not target the same delivery. In addition, multiple developers need to work on the same file or resources and make changes to the same or different parts. To avoid collisions among those changes and enable only completed tasks while keeping not completed, ongoing tasks out, you need to implement a strategy. There are different strategies to solve this problem, but because covering all of them would turn the rest of this book into a Git book, let's focus on only the most popular approach, the branch/merge (pull request) Git flow.

Let's assume you have hired some other developers to work on your project. You are all working on different changes, but as you have learned so far, you do not want to include unfinished changes in your master branch. To protect your development environment, let's start by branching out the current master code.

1. Right click the module and select Git ⇨ Repository ⇨ Branches, as shown in Figure 9-37.

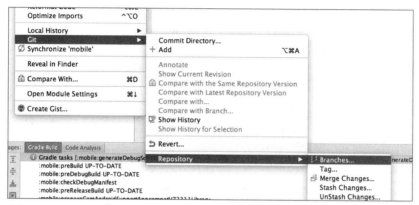

FIGURE 9-37: Branch menu item

This will bring up the Git Branches window. Because you are on the one and only branch in your project, there is nothing else to display.

FIGURE 9-38: New Branch selection

2. Click New Branch, as in Figure 9-38.

Another popup window will ask the branch name, as shown in Figure 9-39. Although you can enter any name in this text field, there are generally accepted conventions for naming a branch in Git.

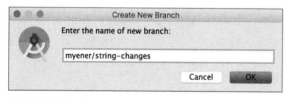

FIGURE 9-39: Branch naming

NOTE *There are several conventions for branch naming that can be accepted as "Good Development Practices." As a best practice, you should keep different branches, such as feature/release/hotfix, instead of directly branching out and merging the branch into the master. Naming your branches according to your task/bug and grouping them with a task or developer name are also good strategies for keeping your Git repository clean and organized.*

1. Start with a simple naming convention for this example. Type your name and a feature name for your new branch, as shown in Figure 9-39.

   You have created and moved your workspace to the new branch. Now it is time to make some changes and commit.

2. Open the `MainActivity` class in the mobile module and find the string `MyAction` in `setAction` in the `onCreate` method.

3. To keep things simple, let's make a small change and delete `My`, leaving the action string as `"Action"`, as shown in Figure 9-40.

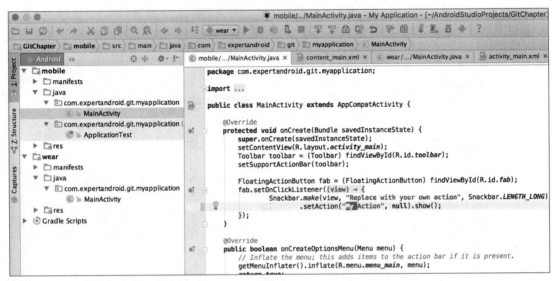

FIGURE 9-40: Simple change in code

Once again, `MainActivity` displays in blue, which indicates that there are some uncommitted changes, as shown in Figure 9-41.

FIGURE 9-41: Uncommitted changes indicated in Android Studio

4. Now select Commit File under Git, as shown in Figure 9-42.

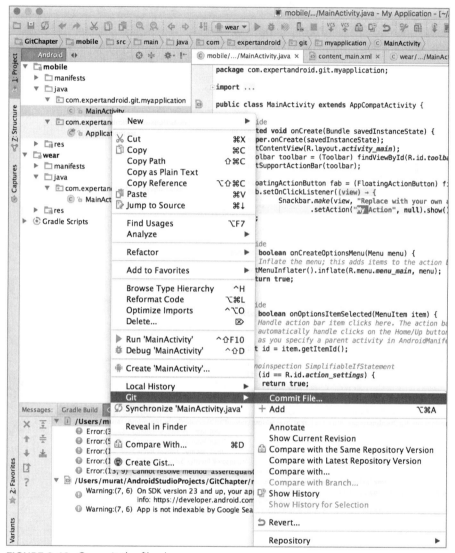

FIGURE 9-42: Commit the file changes

Now let's focus on some other properties and options of the commit window shown in Figure 9-43.

The following list describes the eight icons on top of the file tree:

➤ **Show Diff**—This is one of the most important features of the commit window. This option displays the Diff between the last commit and the current commit candidate. This tool helps you to review what has changed, as shown in Figure 9-44.

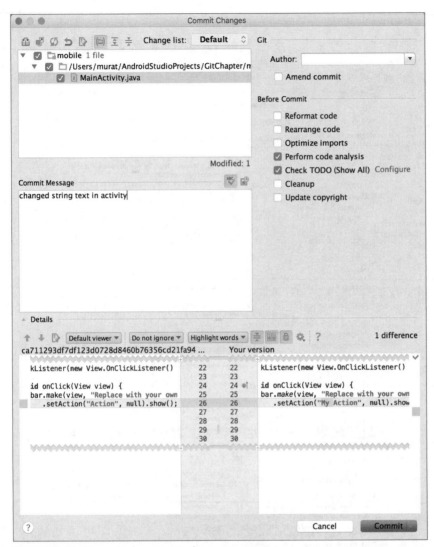

FIGURE 9-43: Commit changes window

➤ **Move to Another Changelist**—Moves current changes into another changelist. Changelists are a group of files with changes. By default, every change is preserved in the default change-list; however, with this option, you can move these changes to another changelist for better and more logical organization, as shown in Figure 9-45.

```
● ● ●                        MainActivity.java (/Users/murat/AndroidStudioProjects/GitChapter/mobile/src/main/java/com/expertandroid/git/myapplication)
↑ ↓ ⏏  Default viewer ▾   Do not ignore ▾   Highlight words ▾  ÷ ▦ ⚲   ☑ Include into commit ⚙  ?                                              1 difference
ca711293df7df123d0728d8460b76356cd21fa94 (Read–only)                              11   11      Your version                                          ˅
                                                                                  12   12
    public class MainActivity extends AppCompatActivity {                         13   13      public class MainActivity extends AppCompatActivity {
                                                                                  14   14
        @Override                                                                 15   15 ⚫        @Override
        protected void onCreate(Bundle savedInstanceState) {                      16   16          protected void onCreate(Bundle savedInstanceState) {
            super.onCreate(savedInstanceState);                                   17   17              super.onCreate(savedInstanceState);
            setContentView(R.layout.activity_main);                               18   18              setContentView(R.layout.activity_main);
            Toolbar toolbar = (Toolbar) findViewById(R.id.toolbar);               19   19              Toolbar toolbar = (Toolbar) findViewById(R.id.toolbar);
            setSupportActionBar(toolbar);                                         20   20              setSupportActionBar(toolbar);
                                                                                  21   21
            FloatingActionButton fab = (FloatingActionButton) findViewById(R.id.fab);  22   22          FloatingActionButton fab = (FloatingActionButton) findViewById(R.id.fab);
            fab.setOnClickListener(new View.OnClickListener() {                   23   23              fab.setOnClickListener(new View.OnClickListener() {
                @Override                                                         24   24 ⚫                @Override
                public void onClick(View view) {                                  25   25                  public void onClick(View view) {
                    Snackbar.make(view, "Replace with your own action", Snackbar.LENG » 26  26 ✕             Snackbar.make(view, "Replace with your own action", Snackbar.LENG
                        .setAction("Action", null).show();                        27   27                          .setAction("My Action", null).show();
                }                                                                 28   28                  }
            });                                                                  29   29              });
        }                                                                        30   30          }
                                                                                  31   31
        @Override                                                                 32   32 ⚫        @Override
        public boolean onCreateOptionsMenu(Menu menu) {                          33   33          public boolean onCreateOptionsMenu(Menu menu) {
            // Inflate the menu; this adds items to the action bar if it is present.  34   34     // Inflate the menu; this adds items to the action bar if it is present.
            getMenuInflater().inflate(R.menu.menu_main, menu);                    35   35              getMenuInflater().inflate(R.menu.menu_main, menu);
            return true;                                                          36   36              return true;
        }                                                                        37   37          }
                                                                                  38   38
        @Override                                                                39   39 ⚫        @Override
        public boolean onOptionsItemSelected(MenuItem item) {                    40   40          public boolean onOptionsItemSelected(MenuItem item) {
            // Handle action bar item clicks here. The action bar will            41   41              // Handle action bar item clicks here. The action bar will
            // automatically handle clicks on the Home/Up button, so long         42   42              // automatically handle clicks on the Home/Up button, so long
            // as you specify a parent activity in AndroidManifest.xml.           43   43              // as you specify a parent activity in AndroidManifest.xml.
            int id = item.getItemId();                                           44   44              int id = item.getItemId();
                                                                                  45   45
            //noinspection SimplifiableIfStatement                               46   46              //noinspection SimplifiableIfStatement
            if (id == R.id.action_settings) {                                    47   47              if (id == R.id.action_settings) {
                return true;                                                     48   48                  return true;
            }                                                                    49   49              }
                                                                                  50   50
            return super.onOptionsItemSelected(item);                            51   51              return super.onOptionsItemSelected(item);
        }                                                                        52   52          }
    }                                                                            53   53      }
Commit Message                                                                                                                                    ⚏ 🖼
changed string text in activity
```

FIGURE 9-44: Diff tool output

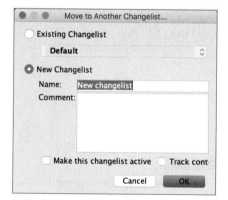

FIGURE 9-45: Changelist window

➤ **Refresh Changes**—Forces a refresh on the contents of the file.

➤ **Revert**—Rolls back the current highlighted change to the previous commit.

➤ **Jump to Source**—Closes the current commit window and focuses on the currently highlighted change in the IDE.

➤ **Group by Directory**—Groups changed files in packages. If this option is not selected, all files will be listed in a flat list that is not dependent on where they are located. In most cases, Group by Directory gives you a better overview of the changed files.

➤ **Expand All**—Expands all packages.

➤ **Collapse All**—Collapses all packages.

That's it for the icons; now let's focus on the Commit window's other features and options.

Click Details on the bottom of Figure 9-43 to expand the rest of the commit window. The Details pane offers a fast, simple diff preview, which we previously covered under the Show Diff feature. Similarly, the quick diff highlights differences between the latest version and previous commit side by side.

Now that we've covered every detail in this window, it is time to perform the commit.

1. Write a commit message and click Commit. Let's assume this was everything you wanted to do. Therefore, it is time to merge and bring those changes from the branch to the head.

2. Right-click the project and select Git ⇨ Repository ⇨ Merge Changes, as shown in Figure 9-46.

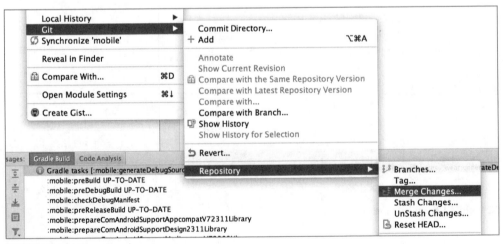

FIGURE 9-46: Merging changes

Next, the Merge Branches window, shown in Figure 9-47, appears. Just under the Git root, which is the root of your project, the current branch and other available branches for merge are displayed.

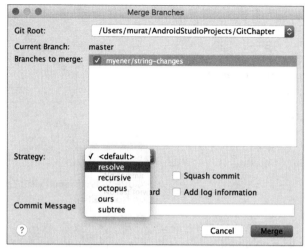

FIGURE 9-47: Branch merge

Because you have only one branch other than the master, it is the only available choice to merge with the master. Just below the branches is a drop-down box to select your merge strategy. Because you currently are the only developer who has performed any change, the strategy you choose is not trivial. However, in real life, you cannot be sure there will not be any conflicts. Go ahead and click Merge.

Well done—you did it! You branched out from the master branch, made a change, and merged the change back. But what really happened? To visualize, we will use an open source tool called GitUp.

> **NOTE** *GitUp is hosted on GitHub. GitUp is a great way to visualize branches, tags, and other Git concepts. GitUp is free and open source and available to download from* `http://gitup.co/`.

We started by branching out from the master, as shown in Figure 9-48. We were on our newly created branch when we took this screenshot. Because it was our focus branch, a red solid line is used. This graph also shows that we have branched out from the master. As we performed our commit and moved back to the master, once again the master has become our focus, as shown in Figure 9-49.

Notice that the dashed branch line is now a solid line, which shows that it has merged back to that master. Because there were no changes in the master when we branched out and merged back in, we merge back at the same location, thus going back and forth on the same line.

FIGURE 9-48: Master branch visualization from GitUp

The last commit/merge was very smooth. Now it is time to take it a step further and create some conflicts. Once again, go back to the beginning of this section and create a new branch. This time, change the `setAction` string to `"Action"` from `"My Action"` in `MainActivity` and perform a commit as you did last time. Switch back to the master branch but do not merge yet.

Now it is time to simulate an already merged change into the master after you branched out. Find the line you just changed and this time change the first string `"Replace"` to `"Replacing"`. Once again, perform a commit with a proper commit message.

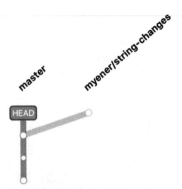

FIGURE 9-49: Branch visualization from GitUp

> **NOTE** *In real life, you should never ever commit to the master in any circumstances. Always branch out, make your commits, test them, and only merge after proper testing/code review.*

You are ready to merge. This time, your branch, which has one new commit, needs to merge into the master but your branch is missing the last change to the master. Because both the master and your branch introduce changes, the changes should be merged so that they don't break or override each other.

> **NOTE** *Overriding may be necessary if one of the changes became obsolete when the other change was committed. Ideally, the obsolete change should have been reverted by the owner of the branch instead of asking for a merge.*

Now select Merge as shown in Figure 9-46. This time, Git will complain about a conflict and inform you that the file merged with a conflict, as shown in Figure 9-50.

You should see both changes that have been added to `MainActivity` but identified with the branch name they belong to.

`<<<<<<< HEAD` marks the beginning of changes in the target branch on which the merge is performed, which is the master in this case. `=======` marks the end of the first change and start of the second change. Finally, `>>>>>>> ` *branchName* marks the target branch that was merged into the master, which is `myener/string-changes` in this case. The warning window offers three options:

➤ **Accept Yours**—Keeps the first change marked with `HEAD` and discards the second.

➤ **Accept Theirs**—Keeps the second change and discards the first one.

➤ **Merge**—Gives you the opportunity to merge both changes.

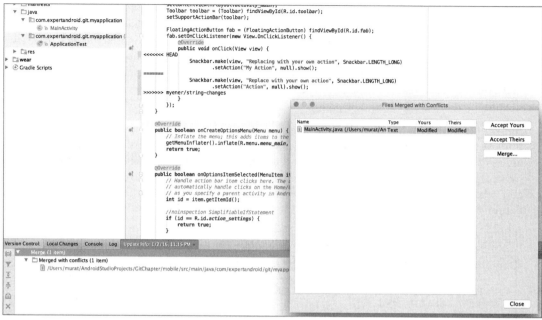

FIGURE 9-50: Merge conflict warning

If those two changes occurred in separate lines, you would not need to perform a manual merge. Instead, Git would automatically bring each new part into the file and create a genuine new version. However, both changes targeted the same line and they are likely to affect each other. In this case, Git allows a manual merge. Click Merge to open the new merge window shown in Figure 9-51.

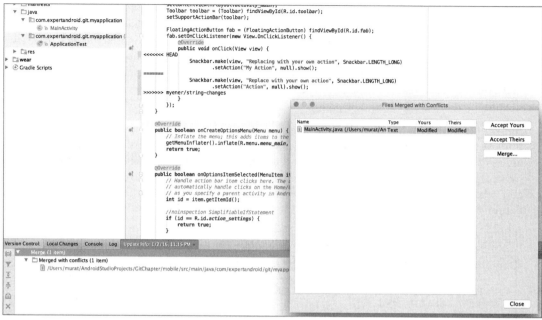

FIGURE 9-51: Merge Revisions window

This window consists of three panes. The left pane shows the current branch you are on, which is the part marked with HEAD in the commit. The right pane displays the change in the branch, which you want to merge with the master. Finally, the middle pane shows a proposed merge from both parties. Git usually does a smart job merging conflicts. There are four buttons at the bottom of the screen: You can choose to Accept Left and discard the change in right pane, choose to Accept Right and override the change in the left pane, click Abort to not perform anything now, or click to Accept the proposed merge in the middle pane.

But wait a minute. What if you are not happy with the proposed change because it doesn't reflect both of the changes because of the conflict between them? Well, if conflicts occur, you can always solve them manually. To help with that, the middle pane is editable. Click the red highlighted part and change Replace to Replacing and My Action to Action. You are even free to add new changes. Let's add a comment to remind you that you made the manual merge shown in Figure 9-52.

FIGURE 9-52: Manual merge

As you finish typing, a green popup will appear with a link to save changes and finish merging, as shown in Figure 9-53.

```
import android.support.v7.widget.Toolbar;
import andro
import andro      All changes have been processed.
import andro      Save changes and finish merging

public class MainActivity extends AppCompatActivity {
```

FIGURE 9-53: Saving changes notification

Because the result of the merge is actually another change, you need to commit this change in your current branch. The commit message automatically included the merge and conflict info, as shown in Figure 9-54. We strongly suggest that you keep this message because it is a standard way to understand what has happened.

Let's see what happened. Once again, we use GitUp to visualize what happened, as shown in Figure 9-55. We originally created a new branch from the master named myener/string-changes. Then we performed a commit on the branch. Meanwhile, a separate change was also committed on the master, as indicated by a small white dot on the red line after we branched out. Finally, we merged those two changes, graphically shown as our branch brought back to the master.

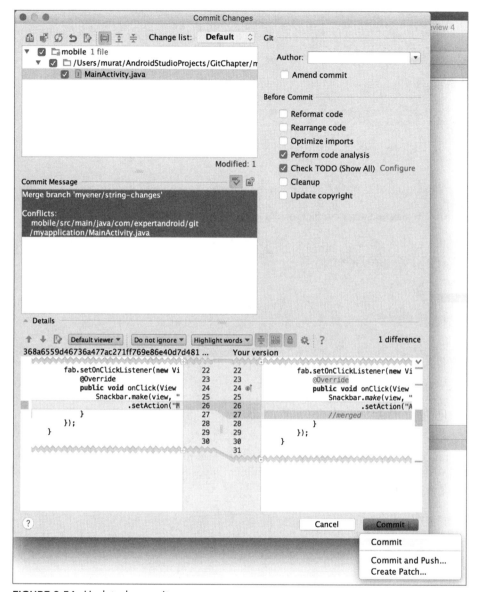

FIGURE 9-54: Updated commit message

As a final step, let's add a remote to our local repository as you previously did from the command line and GitHub client. This time, select Push from the Git menu. Because you have not yet declared a remote repository, Android Studio will display the Define Remote option in the Push Commits window, as shown in Figure 9-56.

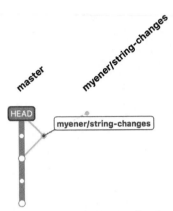

FIGURE 9-55: GitUp visualization after conflict resolution

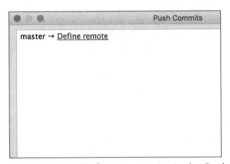

FIGURE 9-56: Defining a remote in the Push Commit window

Clicking the Define remote link opens a new window that asks for the URL of your remote repository, as shown in Figure 9-57. Because you created a local repository in Android Studio, you need to register a new repository on GitHub and enter the URL.

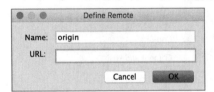

FIGURE 9-57: Remote repository definition

Once you enter the URL, Android Studio will display all commits included in this push, as shown in Figure 9-58. Clicking Push will publish all your changes to the remote repository.

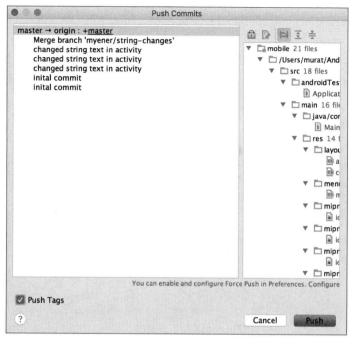

FIGURE 9-58: Push commit summary

SUMMARY

This chapter covered version control systems and focused on the most popular one, Git. You started by creating a Git repository and performing commits from the command line. Next, the chapter covered the popular Git repository, GitHub, and showed you how to use GitHub Desktop Client.

Finally, the chapter moved to Android Studio to create a new Android project. We covered how to create branches and work on them. We also covered Git flow by branching out and merging the changes back to the master branch as well as showing you how to manage conflicts and resolve them.

10

Continuous Integration

WHAT'S IN THIS CHAPTER?

- ➤ Installing Jenkins
- ➤ Configuring Jenkins plugins
- ➤ Integrating Android projects with Jenkins
- ➤ Release management

The previous chapters covered how to manage dependencies, testing, and the Gradle build system, which are crucial pieces of the development life cycle. Those pieces are manually used and triggered. In this chapter, we cover continuous integration (CI) servers, which act as the cement between all other processes and convert them into an automated life cycle.

In this chapter you will learn more about CI and why you need it. You will also download and install your own CI server. Finally you learn how to set up a build job from a Git repository, how to trigger a build cycle on every commit, and how to publish your app automatically to Google Play.

WHAT IS CONTINUOUS INTEGRATION?

> *An important part of any software development process is getting reliable builds of the software. Despite its importance, we are often surprised when this isn't done. We stress a fully automated and reproducible build, including testing, that runs many times a day. This allows each developer to integrate daily thus reducing integration problems.*
>
> —Martin Fowler and Matthew Foemmel, "Continuous Integration"
> (http://martinfowler.com/articles/originalContinuousIntegration.html)

Every software project consists of libraries, modules, and classes that need to integrate with each other. Keeping the integration stable while each piece of integration is subject to change can become a very expensive and time-consuming task. Each introduced change may break another piece of coding integrated with the changed component. Having proper test coverage technically helps to detect such problems. If tests are not consistently and automatically run, the stability of the code depends on how frequently tests are manually executed by developers.

The cost of fixing defects increases proportionally with the length of time it takes to discover them because other modules or systems may also start using the buggy code. Having few and late commit and manual build processes increases the impact of each defect. To have stable projects and builds, you need to minimize human interaction and error as much as possible and automate every eligible piece, including tests and builds.

In a CI system, builds can be broken for several reasons: Tests fail, a component works with some part of the project but fails with the rest, compilation fails, or code quality metrics do not match standards.

When a continuous build system is in action, email(s) with the error log details will be sent to anyone involved in a broken build. Because builds fail almost instantly after commits, CI systems immediately reveal problems and make them visible to everyone.

INTEGRATING ANDROID PROJECTS WITH A CONTINUOUS INTEGRATION SERVER

CI servers are very flexible and easy to integrate and can handle Android projects that use make files, Maven, or Gradle. You need to choose one of those to fully integrate your project with a CI server. This chapter focuses on Gradle, but you may prefer to choose make files or Maven.

Version control systems are another crucial part of the CI process. Each code commit triggers a build process that results in compilation, running tests, and packaging the app on the CI server. We covered the Git version control system in Chapter 9; in this chapter, we focus on integrating a Git project with your CI server.

You need a CI server to do the heavy lifting. Available CI alternatives include Hudson, Jenkins, and Bamboo. Bamboo is a commercial CI server from Atlassian. Hudson and Jenkins are open source, free CI servers used widely in open source and corporate projects and are derived from the same code base.

> **NOTE** *Jenkins is a project built on the original Hudson code base. After Oracle took control of Hudson, developers decided to continue the project under the name Jenkins and to move the project to GitHub. Later, Oracle decided to move the Hudson code base to the Eclipse infrastructure. Both projects are still very similar and support the same plugins up to version 1.395.*

Throughout this chapter, we focus on Jenkins; however, as we mentioned before, you may prefer to use Hudson, which is similar. The Jenkins distribution can be downloaded from `https://jenkins.io/index.html`. At the time of this writing, the latest stable version of Jenkins is 1.654 and Jenkins 2.0 is not yet available.

INSTALLING JENKINS

Jenkins can be downloaded either as a plain WAR file or as an application installer that bundles a web server to run Jenkins. If you already have a Java web server available and running on the target computer, you may prefer to download the WAR file. This section focuses on installing the bundle, which is very straightforward, and the bundle installs with no configuration.

1. Download the Jenkins installer from `https://jenkins.io/index.html`, as shown in Figure 10-1.

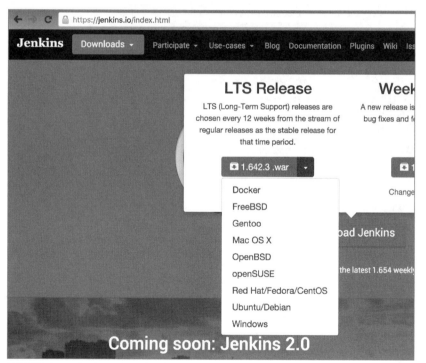

FIGURE 10-1: Jenkins download page

2. If you are running Debian/Ubuntu, open the terminal and type the following commands:

```
wget -q -O - http://pkg.jenkins-ci.org/debian-stable/jenkins-ci.org.key |
sudo apt-key add -
```

```
deb http://pkg.jenkins-ci.org/debian-stable binary/
sudo apt-get update
sudo apt-get install Jenkins
```

If you are using Mac OS X or Windows 10, just click on the installer.

The following steps are identical for each operating system.

3.   Click Continue at the Introduction step, as shown in Figure 10-2.

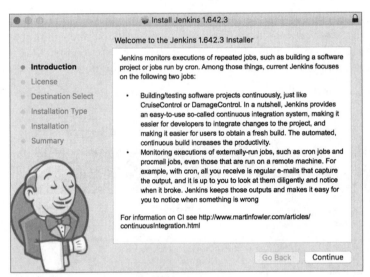

FIGURE 10-2: Jenkins installation window

Jenkins comes with the MIT license.

4.   Click Continue to proceed, as in Figure 10-3.

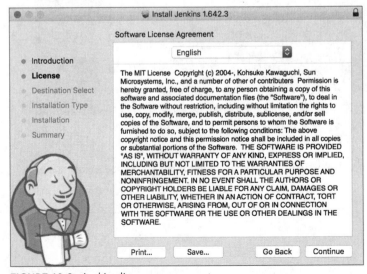

FIGURE 10-3: Jenkins license agreement

5. Select the destination folder where you want to install Jenkins, as shown in Figure 10-4.

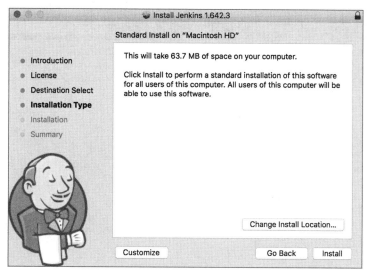

FIGURE 10-4: Jenkins installation directory selection

An information window like the one shown in Figure 10-5 displays when the installation is complete.

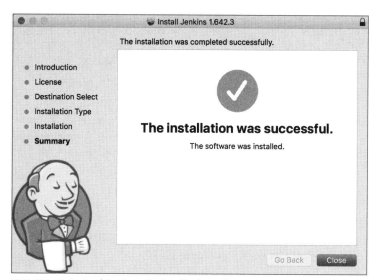

FIGURE 10-5: Jenkins installation final window

After you click Close, the installer will open a browser window pointing to `localhost:8080`, as shown in Figure 10-6. If you have other applications or servers already using 8080, Jenkins might use another port.

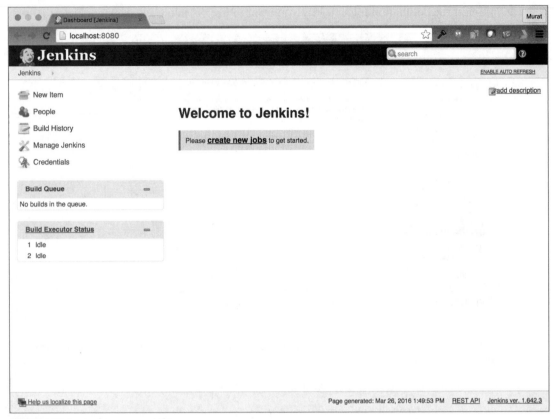

FIGURE 10-6: Jenkins server URL

You have installed Jenkins and it is up and running.

Jenkins relies on plugins to integrate with different setups, project types, and properties. Because you will be using Jenkins for Android projects, you need to install several plugins that differentiate an Android project from a standard Maven-based Java project.

1. Click Manage Jenkins in the left pane, as shown in Figure 10-7.

2. Click Manage Plugins to see the list of available and installed plugins for Jenkins, as shown in Figure 10-8.

3. Select the Available tab, as shown in Figure 10-9, and select Gradle plugin, Git plugin, and GitHub plugin from the search results, as shown in Figure 10-10.

FIGURE 10-7: Accessing Jenkins from a browser

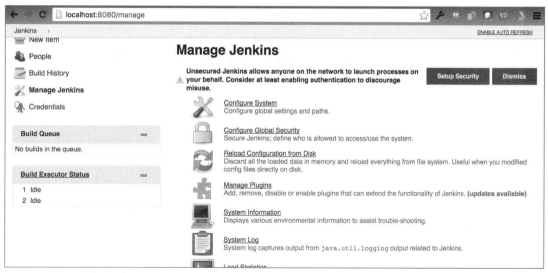

FIGURE 10-8: Jenkins plugins

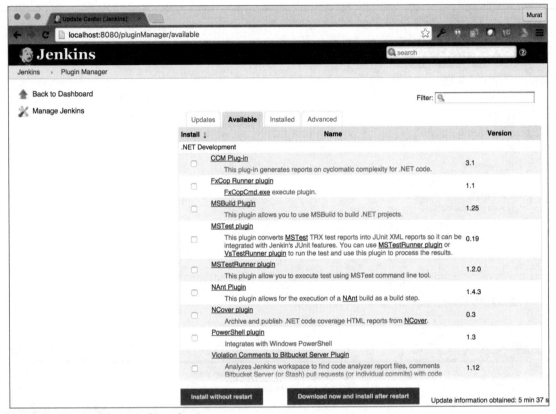

FIGURE 10-9: Searching Jenkins plugins

4. When you are done with the selections, click the Download now and install after restart link to start downloading, as shown in Figure 10-11. The process will continue when you restart Jenkins, as shown in Figure 10-12.

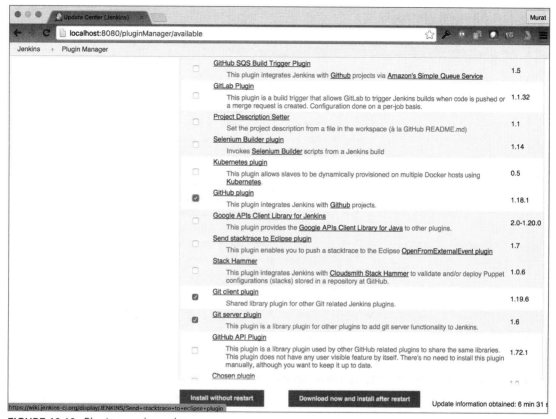

FIGURE 10-10: Plugin search results

Now that you have finished installing Gradle and Git plugins, you can set up a build job to start continuous integration.

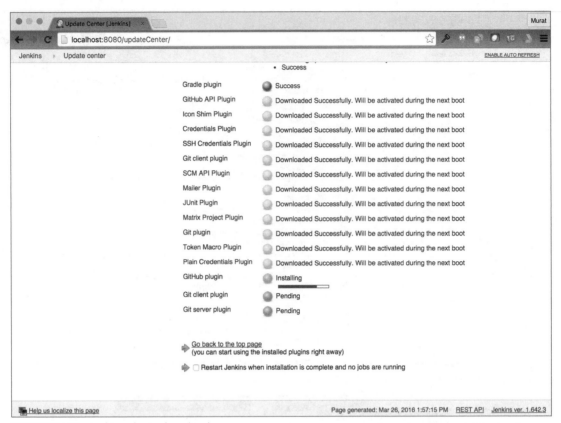

FIGURE 10-11: Jenkins plugin download

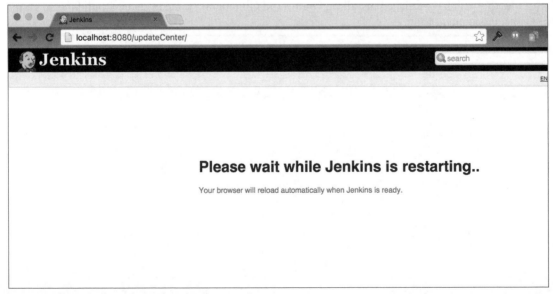

FIGURE 10-12: Jenkins plugin download progress

CREATING BUILD JOBS

Because you have a fresh Jenkins installation with no build jobs yet, Jenkins displays a "create new jobs" option just below the welcome message, as shown in Figure 10-13.

FIGURE 10-13: Creating build jobs

To demonstrate a full build by running tests, generating reports, and building APKs, we need a full project to integrate with our CI server. For this purpose, we will fork the Google I/O 2014 schedule app.

1.   Visit the repository at `https://github.com/kevinmcdonagh/iosched` and click Fork, as shown in Figure 10-14.

FIGURE 10-14: Sample application fork

GitHub will clone the repository into your GitHub account. Now you are ready to create your first build job.

2. Click create new jobs (refer to Figure 10-13) and type a build name, as shown in Figure 10-15.

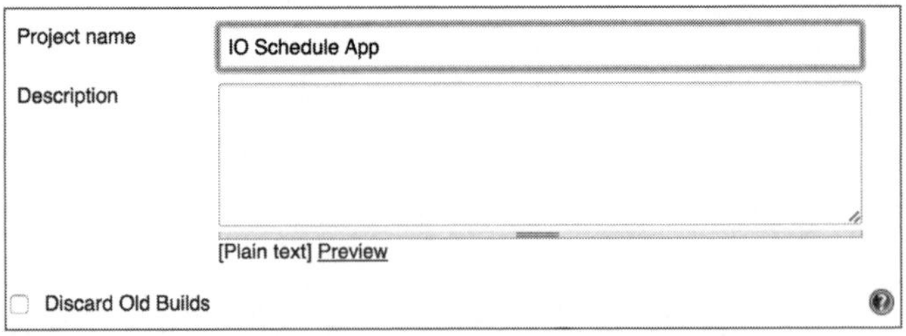

Project name	IO Schedule App
Description	
	[Plain text] Preview
☐ Discard Old Builds	

FIGURE 10-15: Build job for the application fork sample

3. Next, select the GitHub project option and paste your GitHub project URL. Selecting this option is not mandatory because Jenkins can integrate into any Git repository. GitHub acts as a standard GitHub repository, but it lets Jenkins access GitHub-specific metadata and properties, as shown in Figure 10-16.

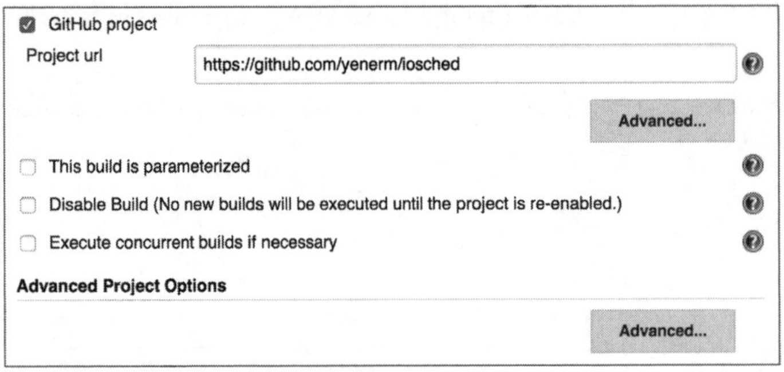

☑ GitHub project
Project url   https://github.com/yenerm/iosched   ⑦

Advanced...

☐ This build is parameterized   ⑦
☐ Disable Build (No new builds will be executed until the project is re-enabled.)   ⑦
☐ Execute concurrent builds if necessary   ⑦

Advanced Project Options

Advanced...

FIGURE 10-16: Jenkins access to GitHub

4. Select a source code management option (Git in our case). Type in your repository URL, as shown in Figure 10-17.

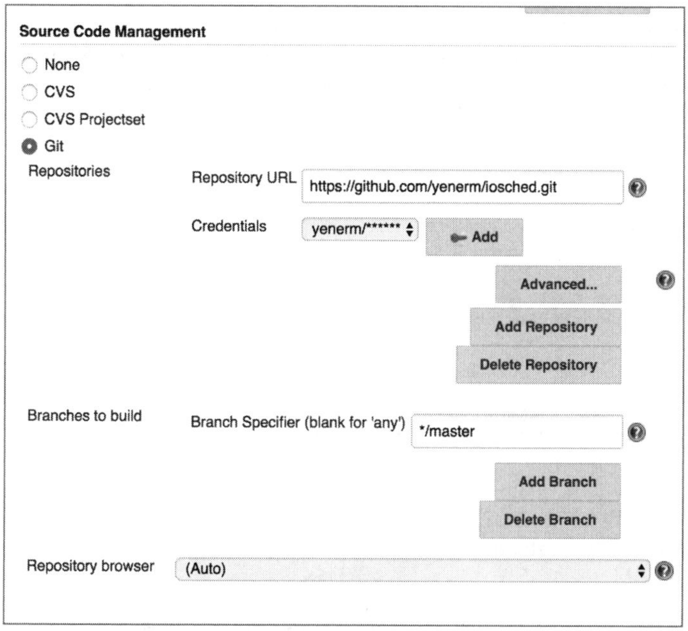

FIGURE 10-17: Repository initialization

5. Click the Add button next to Credentials and add your username and password, as shown in Figure 10-18.

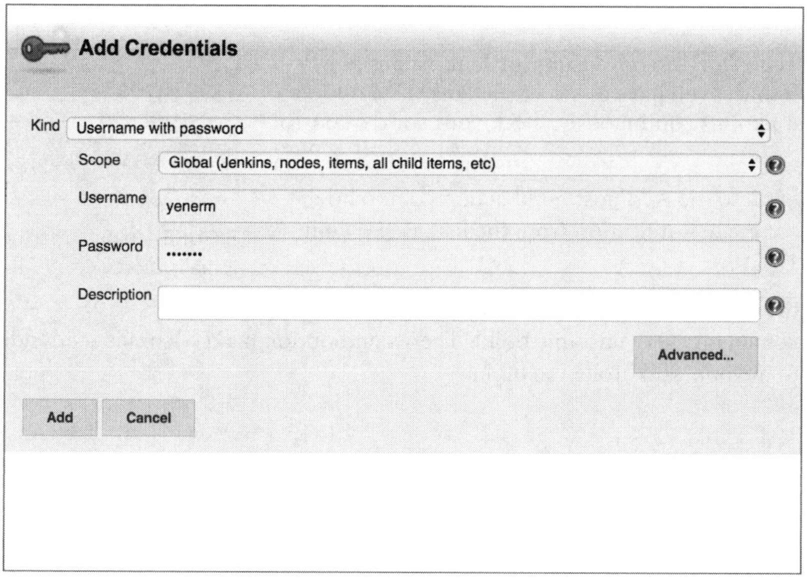

FIGURE 10-18: Credentials for Jenkins

Below the repository properties (refer to Figure 10-17), you can choose which branches to build. Jenkins allows you to build any branch and helps to oversee any integration and stability problems from the outset. For our purposes, we will continue with the master branch.

Build Triggers is another useful setting to control builds. The first option, Build after other projects are built, is used to create a build dependency to another project's build cycle. This is a very useful setting when your build system relies on another library or API that is subject to change. The second option, Build periodically, determines the frequency of periodic builds, as shown in Figure 10-19.

Build Triggers

☐ Build after other projects are built

☑ Build periodically

Schedule

```
H H(0-7) * * *
```

Would last have run at Saturday, March 26, 2016 7:34:05 AM PDT;
would next run at Sunday, March 27, 2016 7:34:05 AM PDT.

☑ Build when a change is pushed to GitHub

☐ Poll SCM

FIGURE 10-19: Build frequency selection

Any desired frequency can be declared in years, months, days, hours, and minutes. Because nightly builds are strongly encouraged in a CI cycle, add a daily build, which would happen between 12 p.m. and 7 a.m. For more information on custom schedules, click the blue question mark next to schedule box.

As a general rule, you want a build to be triggered immediately after every commit to see if the change has broken the build, so select the Build when a change is pushed to the GitHub option.

If your repository does not support Jenkins but publishes changes, you have to select the Poll SCM option to make Jenkins continuously check your source control system.

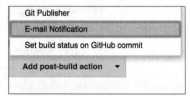

FIGURE 10-20: Email notification initialization

6. Scroll down to the Add post build action combo box and select Email notification from the list, as shown in Figure 10-20.

7. Check the options in Figure 10-21 to allow Jenkins to send an email on every unstable build. The second option makes Jenkins send individual emails to anyone who broke the build.

FIGURE 10-21: Email notification to user

8. Click Apply and go back to the Project View. Click the Build Now option in the left pane, as shown in Figure 10-22.

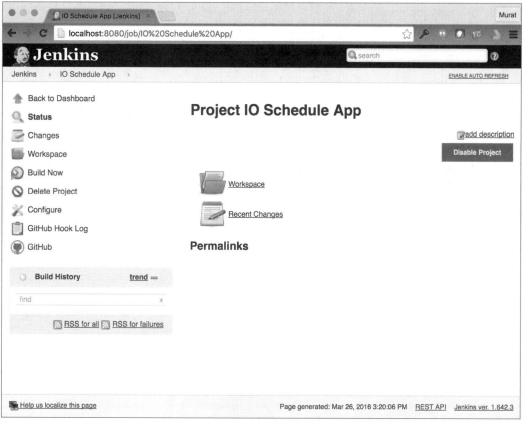

FIGURE 10-22: Building the project

Jenkins will schedule a build and will execute the build process when it is idle, as shown in Figure 10-23.

FIGURE 10-23: Schedule for the build

9. Click the build number to display build properties, as shown in Figure 10-24.

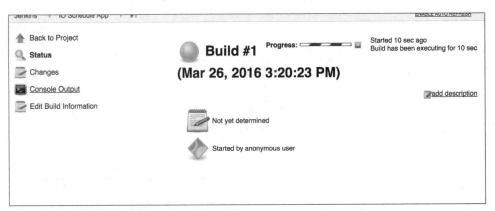

FIGURE 10-24: Build properties by build number

10. Click the Console Output option to view the build messages and log shown in Figure 10-25.

Congratulations, you have just completed your first successful build! Go back to the Jenkins dashboard to view your project status, which should look like Figure 10-26.

FIGURE 10-25: Console output of the build process

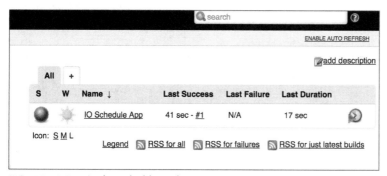

FIGURE 10-26: Jenkins dashboard

The sun icon shown in the W column in Figure 10-27 represents the status of the project. Because the build is successful and you don't have a failed build, everything is sunny. You may see cloudy or even stormy icons depending on the stability of your build.

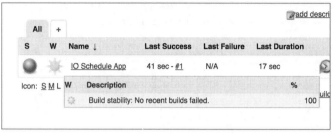

FIGURE 10-27: Status of the build in Jenkins

RELEASE MANAGEMENT

You have integrated your project into Jenkins, created a build schedule, and even had a successful build. However, you are still not utilizing Jenkins's Android-specific capabilities.

1. Go to the Plugin Manager page, select the Available tab, and search for "android," as shown in Figure 10-28.

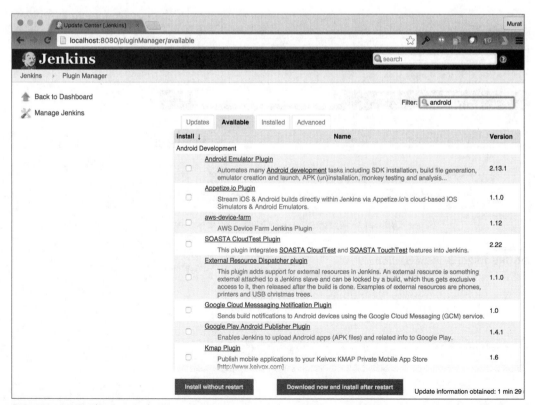

FIGURE 10-28: Plugin search

2. Add the following plugins from search results. These help you to fully utilize your builds.

➤ **Android Emulator Plugin**—Launches an Android emulator in order to run tests.

➤ **Android Lint Plugin**—Generates and displays Lint reports for Android.

➤ **Google Play Android Publisher Plugin** *(Optional)*—Lets you publish your signed APK after a successful build.

➤ **Google Cloud Messaging Notification Plugin** *(Optional)*—Helps test GCM code.

3. Click Download now and install after restart. Jenkins should start downloading the selected plugins and display a screen similar to Figure 10-29.

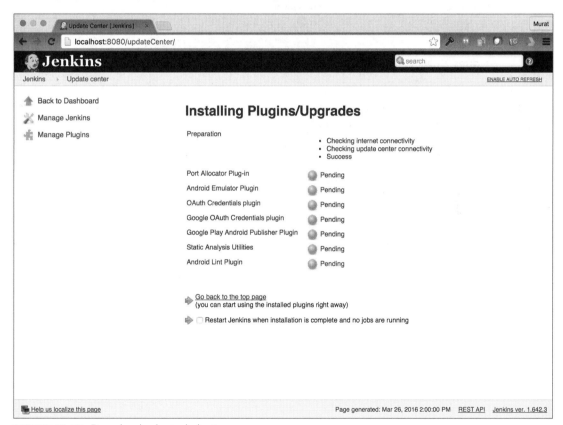

FIGURE 10-29: Download selected plugins

Once the download is finished and Jenkins restarts itself, you can start configuring the new plugins.

4. Select the build job and click Configure, as in Figure 10-30.

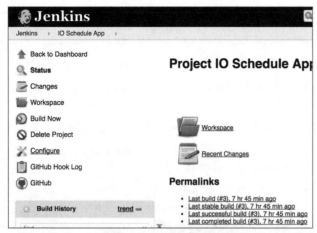

FIGURE 10-30: Configuring a Jenkins build job

5. Scroll down to the Build Environment group and select the Run an Android emulator during build option, as shown in Figure 10-31.

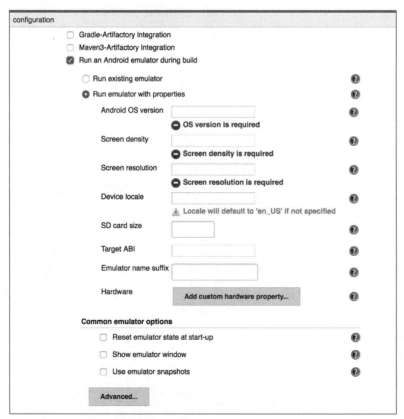

FIGURE 10-31: Running on Android emulator

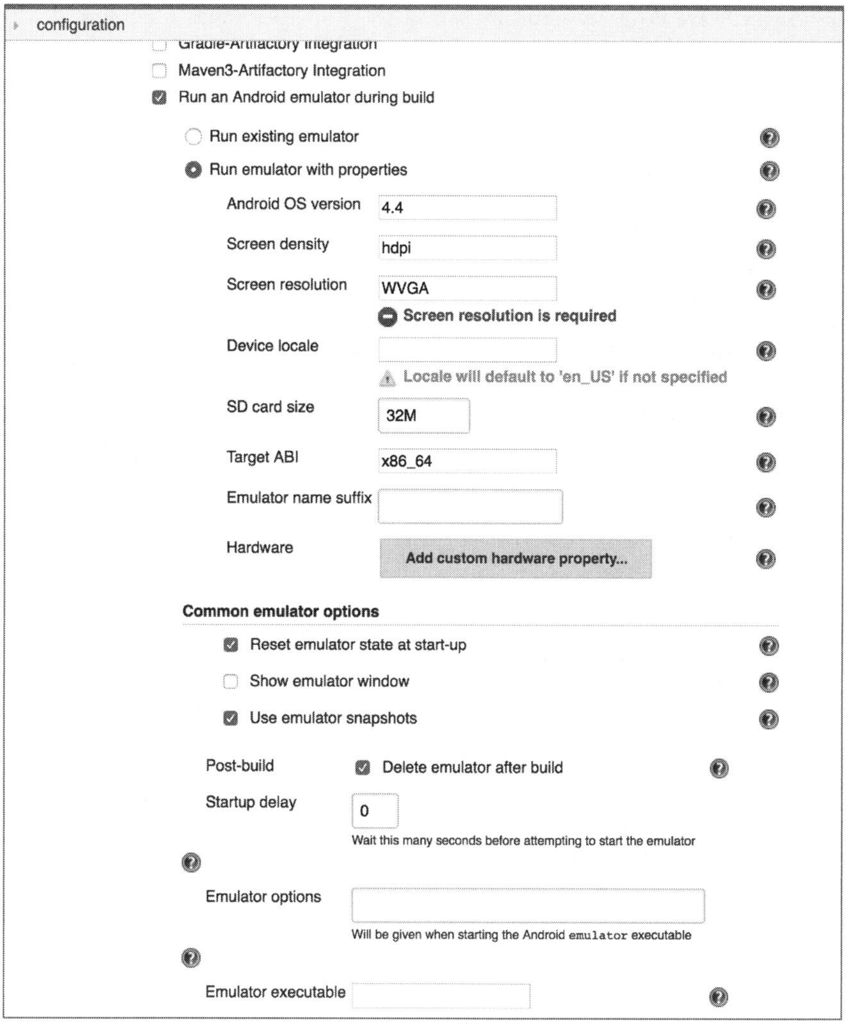

FIGURE 10-32: Emulator configuration

The emulator has the following properties as shown in Figure 10-32:

➤ **Android OS version**—Version of Android operating system.

➤ **Screen density**—Screen density of the emulator device, such as mdpi, hdpi, and so on.

➤ **Screen resolution**—Resolution of screen in width and height or name, such as WVGA.

➤ **Device Locale**—Locale for the device. The en_US locale will be used if left empty.

➤ **SD card size**—Size of the SD card in megabytes or gigabytes.

➤ **Target ABI**—Target architecture such as armeabi, x68, and so on.

➤ **Custom Hardware Property**—Specific hardware properties, such as hw.gps, hw.touchScreen, and so on.

➤ **Reset Emulator State at startup**—Starts a clean emulator without any leftover information from previous runs.

➤ **Show emulator window**—Displays the emulator window.

➤ **Use emulator snapshots**—Uses snapshots for faster startup and initialization of the emulator.

➤ **Startup delay**—Waits a specific time before starting the emulator.

6. If you have downloaded the optional Google Play Android Publisher plugin, scroll down to Build, click Add Build Step, and select Move Android APKs to a different release track. This option will enable Jenkins to publish new builds to the Play store on your behalf. As you can see in Figure 10-33, the Google Play account is not yet configured if you have a fresh Jenkins install.

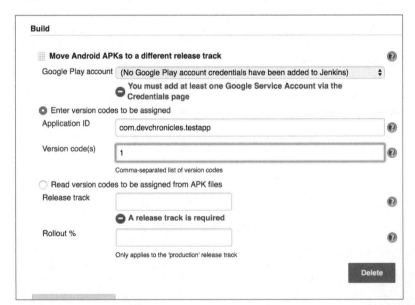

FIGURE 10-33: Google Play account configuration for Jenkins

7. To configure your Google Play account credentials, click Credentials from the left pane, as shown in Figure 10-34.

8. Select the Google Service Account from private key option from the options shown in Figure 10-35.

9. Add your project name and JSON key, which you can download from the Google Developer Console, as shown in Figure 10-36.

FIGURE 10-34: Google Play account credentials configuration

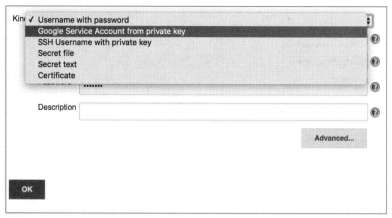

FIGURE 10-35: Google Service Account private key

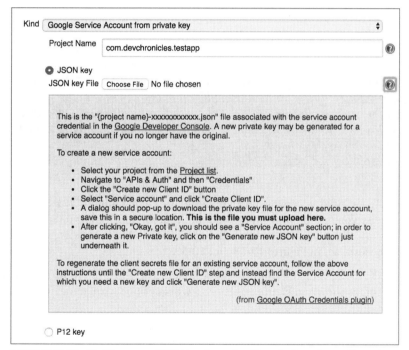

FIGURE 10-36: Getting your JSON key from the Google Developer Console

Now you can go back to build job settings to complete the Move Android APKs to a different release track option.

10. Select the Google Play account you have just created. Enter the Application ID and Version code(s). Choose a release track (alpha, beta, or production) and finally the rollout percentage, as shown in Figure 10-37.

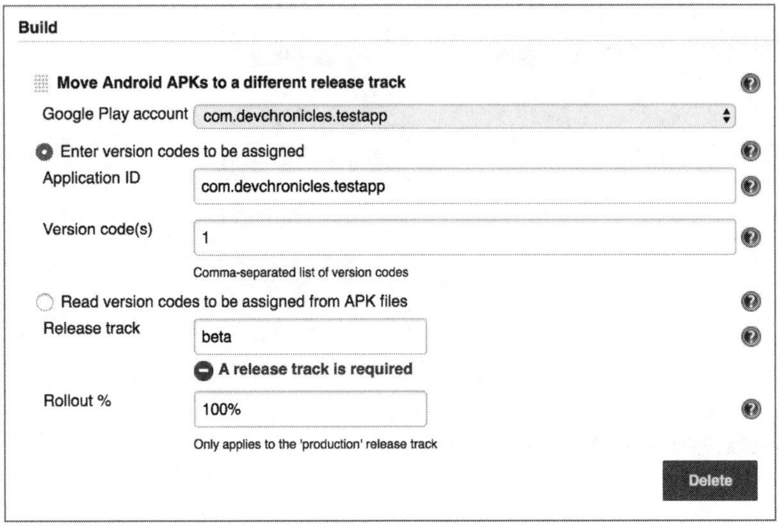

FIGURE 10-37: Release configuration

> **NOTE** *Although auto deployment to production is a tempting option, it should be used with caution. When an APK is published to a production track, it is updated automatically on most devices. Even with very well tested APKs, it is wise to deploy the APK into a beta track, manually test it, and promote the APK for the production track by hand.*

You have configured your build, so now you can go back and trigger a build and start watching the console output. Because you haven't previously used the Android-specific capabilities of Jenkins, the first build will automatically download and install Android SDK, Android tools, the emulator, and the emulator images, as shown in Figures 10-38 and 10-39.

Finally, when the SDK, tools, emulator, and emulator images are ready, the Android emulator plugin will create an emulator, as shown in Figure 10-40, and will continue with launching, as shown in Figure 10-41.

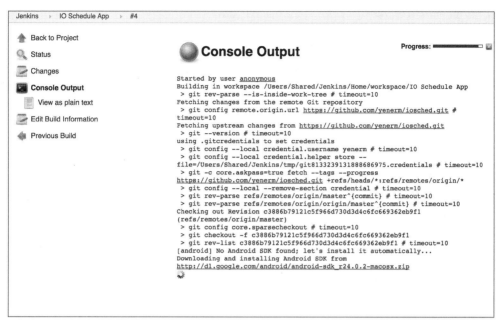

FIGURE 10-38: Build progress to install Android SDK and tools

```
(or rights in favor of) them. Other than this, no other person or company shall
be third party beneficiaries to the License Agreement.

14.5 EXPORT RESTRICTIONS. THE PREVIEW IS SUBJECT TO UNITED STATES EXPORT LAWS
AND REGULATIONS. YOU MUST COMPLY WITH ALL DOMESTIC AND INTERNATIONAL EXPORT LAWS
AND REGULATIONS THAT APPLY TO THE PREVIEW. THESE LAWS INCLUDE RESTRICTIONS ON
DESTINATIONS, END USERS AND END USE.

14.6 The License Agreement may not be assigned or transferred by you without the
prior written approval of Google, and any attempted assignment without such
approval will be void. You shall not delegate your responsibilities or
obligations under the License Agreement without the prior written approval of
Google.

14.7 The License Agreement, and your relationship with Google under the License
Agreement, shall be governed by the laws of the State of California without
regard to its conflict of laws provisions. You and Google agree to submit to the
exclusive jurisdiction of the courts located within the county of Santa Clara,
California to resolve any legal matter arising from the License Agreement.
Notwithstanding this, you agree that Google shall still be allowed to apply for
injunctive remedies (or an equivalent type of urgent legal relief) in any
jurisdiction.

June 2014.

Do you accept the license 'android-sdk-preview-license-52d11cd2' [y/n]:
Installing Archives:
  Preparing to install archives
  Downloading Android SDK Platform-tools, revision 23.0.1
      (3%, 41 KiB/s, 56 seconds left)
      (5%, 36 KiB/s, 63 seconds left)
      (11%, 63 KiB/s, 34 seconds left)
      (17%, 79 KiB/s, 25 seconds left)
      (24%, 90 KiB/s, 20 seconds left)
      (26%, 85 KiB/s, 20 seconds left)
      (28%, 80 KiB/s, 21 seconds left)
      (32%, 78 KiB/s, 21 seconds left)
      (34%, 77 KiB/s, 20 seconds left)
      (42%, 86 KiB/s, 16 seconds left)
      (46%, 88 KiB/s, 14 seconds left)
      (59%, 104 KiB/s, 9 seconds left)
      (67%, 110 KiB/s, 7 seconds left)
```

FIGURE 10-39: Build progress to install the emulator

```
   Unzipping SDK Platform Android 4.4.2, API 19, revision 4 (97%)
     Installed SDK Platform Android 4.4.2, API 19, revision 4
   Done. 1 package installed.
[android] Using Android SDK: /Users/Shared/Jenkins/Home/tools/android-sdk
[android] Creating Android AVD: /Users/Shared/Jenkins/.android/avd/hudson_en-
US_240_WVGA_android-19_x86_64.avd
[android] /Users/Shared/Jenkins/Home/tools/android-sdk/tools/android create avd
-f -a -c 32M -s WVGA800 -n hudson_en-US_240_WVGA_android-19_x86_64 -t android-19
--abi x86_64
```

FIGURE 10-40: Jenkins creating the emulator

```
The filesystem is already 51200 (4k) blocks long.  Nothing to do!

Creating filesystem with parameters:
    Size: 209715200
    Block size: 4096
    Blocks per group: 32768
    Inodes per group: 6400
    Inode size: 256
    Journal blocks: 1024
    Label:
    Blocks: 51200
    Block groups: 2
    Reserved block group size: 15
Created filesystem with 11/12800 inodes and 1865/51200 blocks
Creating filesystem with parameters:
    Size: 69206016
    Block size: 4096
    Blocks per group: 32768
    Inodes per group: 4224
    Inode size: 256
    Journal blocks: 1024
    Label:
    Blocks: 16896
    Block groups: 1
    Reserved block group size: 7
Created filesystem with 11/4224 inodes and 1302/16896 blocks
emulator: UpdateChecker: skipped version check
$ /Users/Shared/Jenkins/Home/tools/android-sdk/platform-tools/adb connect
localhost:9735
unable to connect to localhost:9735: Connection refused
[android] Waiting for emulator to finish booting...
$ /Users/Shared/Jenkins/Home/tools/android-sdk/platform-tools/adb -s
localhost:9735 shell getprop init.svc.bootanim
error: device 'localhost:9735' not found
$ /Users/Shared/Jenkins/Home/tools/android-sdk/platform-tools/adb connect
localhost:9735
$ /Users/Shared/Jenkins/Home/tools/android-sdk/platform-tools/adb -s
localhost:9735 shell getprop init.svc.bootanim
error: device 'localhost:9735' not found
$ /Users/Shared/Jenkins/Home/tools/android-sdk/platform-tools/adb connect
localhost:9735
```

FIGURE 10-41: Jenkins launching the emulator

If the Play account credentials are correct, the Google Play plugin will publish your APK after a successful build and test process.

SUMMARY

This chapter discussed the importance of stable and reliable software development. Continuous integration can greatly help minimize bugs and make them visible even in very early stages of application development. CI will also automate the building process, letting developers focus on development tasks.

We covered Jenkins, a widely used and accepted continuous integration server. We focused on how to make Jenkins work with Gradle and Android by making use of the plugins. We covered how to build, test, and publish Android projects by pulling the source from version control and publishing the APK to the Play Store.

Continuous integration servers are the integration point where Gradle and proper test coverage start to shine and where software projects can succeed.

11

Using Android NDK with Android Studio

This chapter focuses on the details of Android NDK and shows you how to build native C/C++ code in Android Studio. Native code is commonly used in Android projects for games and applications, which require high performance face recognition, audio processing, and so on. Although Android NDK is a powerful tool, many Android developers and projects may not need to use it. This chapter does not aim to teach Android NDK from the ground up but focuses instead on how to use Android NDK with the new Android Studio and Gradle.

At the time of this writing, Android NDK integration with Android Studio is still experimental and subject to change. We strongly suggest keeping your tools up-to-date and that you follow the updates to NDK integration if your application relies on NDK.

If you are an Android NDK newbie looking to learn NDK, we suggest you visit `http://developer.android.com/ndk` and follow the tutorials and code samples. NDK might look scary if you are not familiar with C/C++; however, it can unleash the full potential of your device's hardware and native libraries, and is well worth the effort.

INTRODUCTION TO ANDROID NDK

Android NDK is an essential part of Android development that lets developers use C/C++ code from Java via JNI. Although the history of Android dates back to 2003, by the time the SDK was released, iPhone was already the main player in the mobile market with a growing application store. To compete, Android needed fast adoption in the developer community. Thus, Google's decision to promote Java as the main language was wise and worked quite well. Android activities, UI widgets, and APIs are all designed in a way that can be used through Java. Java and the Dalvik VM did a great job of lowering the learning curve but that solution lacked the performance that most games and some apps need more than they need features like garbage collection.

Android NDK, which was actually released several months after Android SDK, addressed such performance concerns with native code that can be loaded from Java code. When it was released, Android NDK relied on command-line tools, unlike Android SDK, which can be compiled and run via Eclipse IDE.

With the release of Android Studio, NDK was once again left out of the official tool set. At Google I/O 2015, Google finally announced Android NDK support in Android Studio.

ANDROID STUDIO NDK INTEGRATION

NDK integration for Android Studio was announced with version 1.3. At the time of this writing, NDK integration with Android Studio is still beta and relies on an experimental version of the Gradle plugin. Before going forward with NDK use cases in Android Studio, let's see how to install NDK for Linux, Windows 10, and Mac OS X.

Android NDK packages can be accessed at `http://developer.android.com/ndk/downloads/index.html`. There you will see the list of packages for Linux, Windows, and Mac OS X.

> **NOTE** *There are common steps for NDK installation on all operating systems. In following sections, if you install NDK using Android Studio, you will be asked to accept the license agreement, as shown in Figure 11-1.*

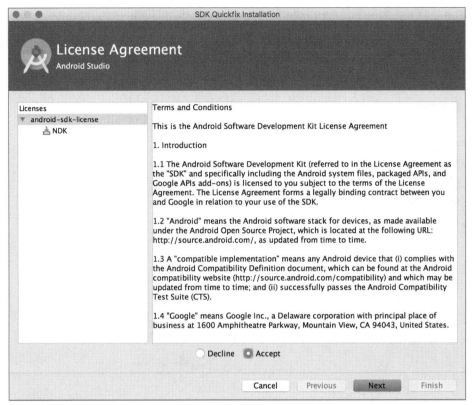

FIGURE 11-1: Android NDK bundle license agreement

Android NDK Installation on Linux

In this section, you install Android NDK on Ubuntu 14.04. Download the Linux 64-bit version, `android-ndk-r11b-linux-x86_64.zip`, from the URL mentioned in the previous section. Next, extract NDK to the Android SDK installation root. for example `/path/to/Android/Sdk/android-ndk-r11b`.

The zip file contains all required binaries to build native Android code.

You can also install Android NDK from Android Studio's SDK Manager. The easiest way to do this when you have an open project is to select the Project Structure option from the File menu to open the Project Structure window. Select SDK Location, as shown in Figure 11-2.

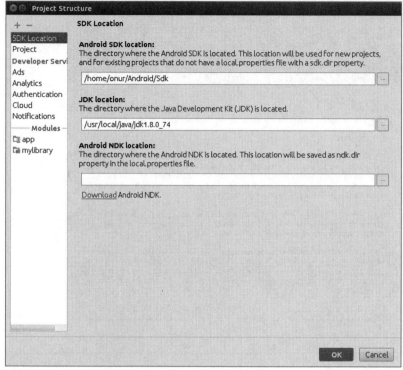

FIGURE 11-2: Project Structure window

You will see that the Android NDK location box is empty. You can give the path to the location where you extracted the NDK package or click the Download Android NDK link to have Android Studio install it for you, as shown in Figure 11-3.

Android Studio will install NDK to the ndk-bundle folder in your SDK path.

Android NDK Installation on Windows 10

There are two ways to install Android NDK for Android Studio on Windows 10. You can install it manually by downloading it from https://developer.android.com/ndk/downloads/index .html or you can install it from Android Studio's Project Structure window.

To install Android NDK from Android Studio, open the Project Structure window, select SDK Location in Android Studio, and click the Download link under Android NDK Location as shown in Figure 11-4. After you accept the license agreement, the Android NDK binaries and libraries will be extracted into the Android SDK ndk-bundle folder.

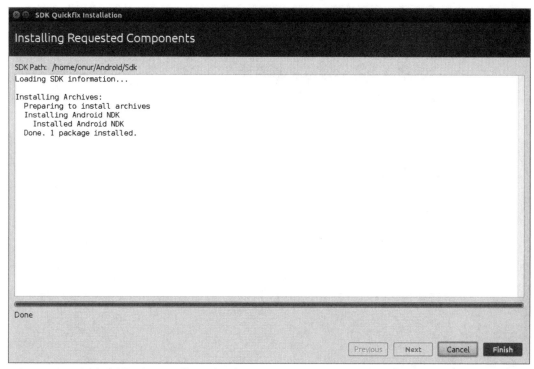

FIGURE 11-3: Android Studio installs Android NDK

No further configuration is needed; you are ready to use Android NDK for your project in Windows 10.

If you want to download Android NDK manually, you can select either the 32-bit or the 64-bit version. Choose version that's appropriate for your Windows 10 machine architecture. In our case we downloaded `android-ndk-r11b-windows-x86_64.zip` and extracted it to a folder.

When you finish extracting the zip file, enter the path to the folder in the Android NDK Location text box as shown in Figure 11-4. Now you are ready to use Android NDK for your Android Studio project.

Android NDK Installation on Mac OS X

Android NDK installation for Mac OS X can be done either by downloading it from `https://developer.android.com/ndk/downloads/index.html` or you can open the Project Structure window to download and extract NDK.

If you choose to install manually, navigate to the URL just mentioned and click the link for the `android-ndk-r11b-darwin-x86_64.zip` file.

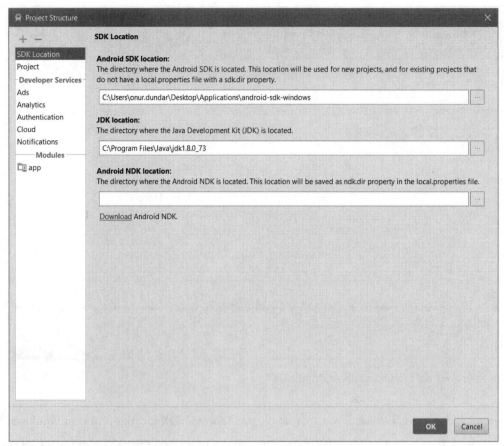

FIGURE 11-4: Android Studio Project Structure window in Windows 10

> **NOTE** *The NDK version can change—it was r11b at the time of this writing.*

Extract the zip file to the folder where you want to keep NDK files. Then open the Project Structure window and enter the folder's path in the Android NDK Location text box. Now your project is ready to use Android NDK.

Alternatively, Android Studio can download and extract Android NDK automatically. Open the Project Structure window and select SDK Location, then click Download in the Android NDK Location section. After clicking Download, accept the license agreement, then click Next to open the Component Installer window shown in Figure 11-5.

When the installation finishes, the last line in this windows will read "Installation of NDK complete," and you can click Finish.

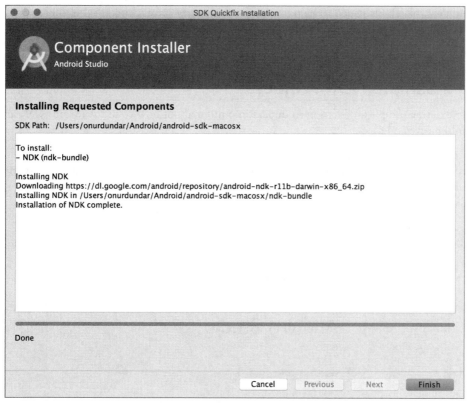

FIGURE 11-5: Android NDK setup completion

Now you can use NDK tools to build native C/C++ code for your application.

ANDROID NDK WITH ANDROID STUDIO PROJECTS

Now that NDK is installed, you can create a new project and start adding your native code to it. (Note that Android Studio does not offer a specific wizard or a template to create NDK projects.)

We first cover some common user preferences, how native applications handle where the native code should be stored, how to create the Gradle file, and so on.

In legacy native application development, Android make files are used to build native modules. These make files were used to define the environment variables and the path of the ndk-build binary to build native C/C++ code. However, when developing in Android Studio, the Gradle build system is used to build native code instead of Android make files.

Native code is usually stored under the jni folder, which is at the same level as the java folder. The output of native library code is a shared library, a .so file. You also need to identify the .so file location according to its compatible architecture. The main folder for libraries should be named jniLibs. Subfolders, which also need to be named so as to identify the architectural elements they hold (such as mips, x86, armeabi, and so on), need to be created under the jniLibs folder.

In the following section, you work on the sample HelloJNI sample.

Importing a Sample NDK Project

Let's open an existing sample application to understand more about how to use Android NDK and Gradle to build native code.

As in earlier chapters, we will import an Android code sample. Open the Welcome to Android Studio window by closing your currently open project; then click Import an Android code sample, as shown in Figure 11-6.

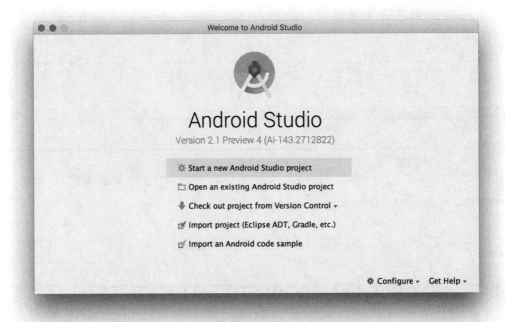

FIGURE 11-6: Importing a sample project from the Welcome to Android Studio window

When the Browse Samples window opens, enter **ndk** in the Select a sample to import box to filter NDK projects, as shown in Figure 11-7.

FIGURE 11-7: Sample NDK projects list

Select the Hello JNI project, which you will use to learn the basics of NDK development with Android Studio. Click Next to open the Configure Sample window shown in Figure 11-8 where you name the sample application and download source code from GitHub.

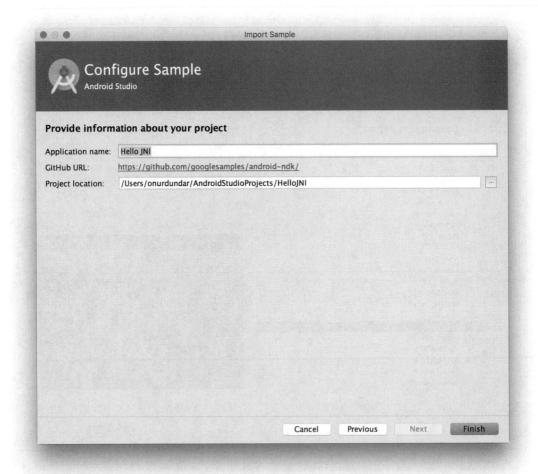

FIGURE 11-8: Sample configuration

Google's code samples for Android Studio are hosted on GitHub; download and import the sample project, as shown in Figure 11-9.

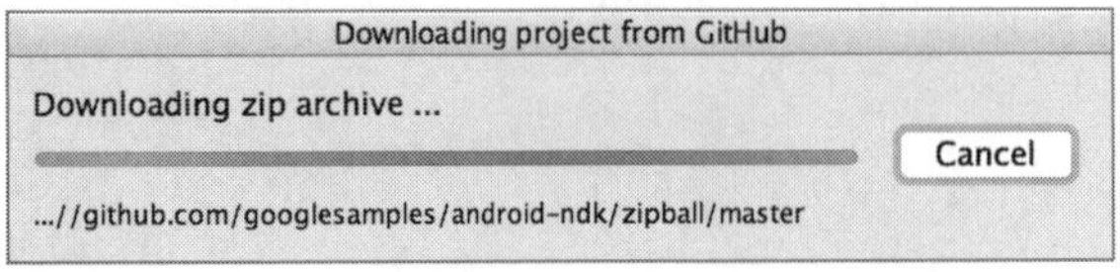

FIGURE 11-9: Downloading the sample project

Once the import is complete, expand the java and jni folders to locate Java and C code for the project, as shown in Figure 11-10.

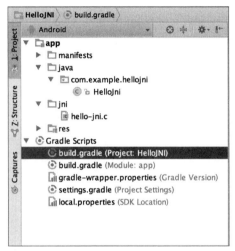

FIGURE 11-10: Project View of the Hello JNI project

Now check the basic integration of the NDK module into the project. Start by expanding the project scope in the build.gradle file and locate the experimental Gradle plugin declaration, which may introduce a different version number than the one used in the previous section.

At the time of this writing, experimental plugin version 0.7.0-alpha1 is used for `classpath`. The version name might be different by the time you read this chapter. The project's Gradle file is shown in the following snippet.

```
buildscript {
    repositories {
        jcenter()
    }
    dependencies {
        classpath 'com.android.tools.build:gradle-experimental:0.7.0-alpha1'
    }
}

allprojects {
    repositories {
        jcenter()
    }
}
```

Next, expand the build.gradle file of the app module. You should notice that the `com.android`
`.model.application` plugin is used instead of the regular `com.android.application` plugin.
You should also see DSL syntax changes such as model and `android.ndk` _moduleName_ as well as

different build flavors to target different platforms and architectures. The jni folder contains the `hello-jni.c` source file, which has been declared as an NDK module in project scope in the build `.gradle` file, as shown in the module's Gradle code in Listing 11-1.

LISTING 11-1: Hello JNI module Gradle script

```
apply plugin: 'com.android.model.application'

model {
    android {
        compileSdkVersion = 23
        buildToolsVersion = "23.0.2"

        defaultConfig.with {
            applicationId = "com.example.hellojni"
            minSdkVersion.apiLevel = 4
            targetSdkVersion.apiLevel = 23
        }
    }

    /*
     * native build settings
     */
    android.ndk {
        moduleName = "hello-jni"
        /*
         * Other ndk flags configurable here are
         * cppFlags.add("-fno-rtti")
         * cppFlags.add("-fno-exceptions")
         * ldLibs.addAll(["android", "log"])
         * stl        = "system"
         */
    }
    android.buildTypes {
        release {
            minifyEnabled = false
            proguardFiles.add(file('proguard-rules.txt'))
        }
    }

}
```

The Hello JNI sample is the most basic use of NDK in Android applications. Its C code returns only a string and will be used in your activity via JNI. Open the Android activity code, shown in Listing 11-2.

LISTING 11-2: JNI call

```
public class HelloJni extends Activity {
    static {
        System.loadLibrary("hello-jni");
```

```
    }
    public native String  stringFromJNI();

    @Override
    public void onCreate(Bundle savedInstanceState)
    {
        super.onCreate(savedInstanceState);

        TextView  tv = new TextView(this);
        tv.setText( stringFromJNI() );
        setContentView(tv);
    }
}
```

This listing demonstrates a simple example but includes all necessary steps to call your C/C++ code from Java. Because your Java code runs on a VM-managed runtime, you need to manually load the non-vm managed code. Static blocks serve well for this purpose because they are initiated even before the constructor and only once for a class.

Java is a type-safe and strongly-typed language, which makes it impossible to see and call the non-vm code even if it is loaded successfully. To be able to call C/C++ code from Java, you would need a placeholder method, which acts as a gateway proxy. Methods, which are marked with the `native` keyword, do not have a method body but share a special naming convention with their C/C++ counterpart; thus, they know which native code to execute when C/C++ codes are accessed.

In the example here, the `stringFromJNI` method returns a string and is called when the activity is created. That's it. If you run the sample application, the activity will call the C code and display the returned string. Try changing the string returned from the C code and see how the changes are reflected to the UI.

Sample NDK Applications

As the list in Figure 11-7 shows, there are multiple sample applications that you can load to learn more about NDK application development. We will give a quick overview of the samples so you can pick one to learn more about the relationship between Android Studio and NDK.

Using Android NDK in an application requires high performance graphics processing, image processing, and audio processing. The following NDK sample applications are mostly focused on showing basic uses of those features with Android NDK. This list of sample applications follows the order shown in Figure 11-7:

➤ **More Teapots**—This sample application uses OpenGLES 3.0 and C++ to draw teapots. It shows the use of the Gradle Experimental Android Plugin and the Android Native Activity.

➤ **Teapot**—This sample application draws a single teapot using the OpenGLES 2.0 API using C++. Its aim is to show how you can use OpenGLES 2.0 with the Gradle Experimental Android Plugin in Android Studio.

➤ **Audio-Echo**—This sample application uses OpenSL ES to create an audio player and recorder. Audio processing can be costly if you require a high-performance, low-power-consumption application. Using a native library such as OpenSL ES helps to create a better sound application. The Audio-Echo sample can be good practice to learn the basics. This sample application also uses the Gradle Experimental Android plugin.

➤ **Bitmap Plasma**—This sample application shows how to use the `Bitmap` class inside NDK to render a plasma effect and open with a JNI interface on an Android device. The Bitmap Plasma application also uses the Gradle Experimental Plugin to build native code under the jni folder.

➤ **Endless Tunnel**—This application is a good entry point to show how to develop a game using the Android native app glue to create native activity. This application is a little more complex than others because it is literally a game with OpenGL Mathematics library support.

➤ **Hello GL2**—This application offers a basic introduction to the GLES 2.0 API, loading a triangle on the native activity. This is a nice entry point for learning NDK instead of the more complex samples listed earlier.

➤ **Hello Third Party**—This application shows how you can load native third-party libraries and include those libraries in your build process using the Gradle Experimental Android plugin. This is also a good entry point for NDK development; the main purpose of NDK development is reusing your previously developed and built native libraries.

➤ **Native Activity**—As the name implies, this application shows how to create a native Android activity.

➤ **Native Plasma**—This application is similar to Bitmap Plasma, but it doesn't use a Java activity. Instead, it uses a native activity to create and show a rendered plasma effect.

➤ **San Angeles**—This sample application might be the most complex because it is a port of an existing C++ application to Android. The aim of this sample is to show how you should port any existing native source code to Android and build it in Android Studio using the Gradle Experimental Android plugin.

➤ **Sensor Graph**—Finally, this sample application shows developers how to access hardware resources with native code. We know that there are Java libraries that give access to accelerometer, gyroscope, and other sensors, but you will definitely need to access hardware resources to make games playable. This sample only reads the values and prints them on the screen.

Migrating an Existing NDK Project

The experimental Gradle plugin, which provides support for NDK, introduces some changes in Gradle DSL. This section teaches you how to use the plugin to migrate an existing NDK project.

1.   Start Android Studio and select New ➪ Import Project from the File menu, as shown in Figure 11-11. Android Studio will migrate your Eclipse project into a Gradle project.

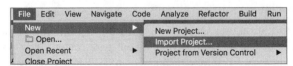

FIGURE 11-11: Importing an NDK project

Once the import is complete, you can move on to making changes on Gradle files.

2. Expand the Gradle Scripts group, which consists of a build.gradle file for the project and a build.gradle file for each module, as shown in Figure 11-12.

FIGURE 11-12: Gradle scripts after importing an NDK project

3. Find the Gradle plugin declaration under the dependencies group. Change this dependency to the bold one in Listing 11-3. Android Studio doesn't load the experimental plugin by default, so you need to change it to the experimental one in order to build native code.

LISTING 11-3: Project Gradle script

```
buildscript {
    repositories {
        jcenter()
    }
    dependencies {
        classpath 'com.android.tools.build:gradle-experimental:0.7.0-alpha1'
        // NOTE: Do not place your application dependencies here; they belong
        // in the individual module build.gradle files
    }
}
```

Changing the declaration as shown tells Android Studio to use the experimental Gradle plugin for your project. Android Studio will detect the change and ask you to perform project sync.

4. Click Sync Now in the upper right, as shown in Figure 11-13.

The build will fail with a message like the one shown in Figure 11-14. Don't worry: You haven't yet completed the necessary changes.

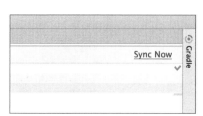

FIGURE 11-13: Syncing the project after changing the plugin

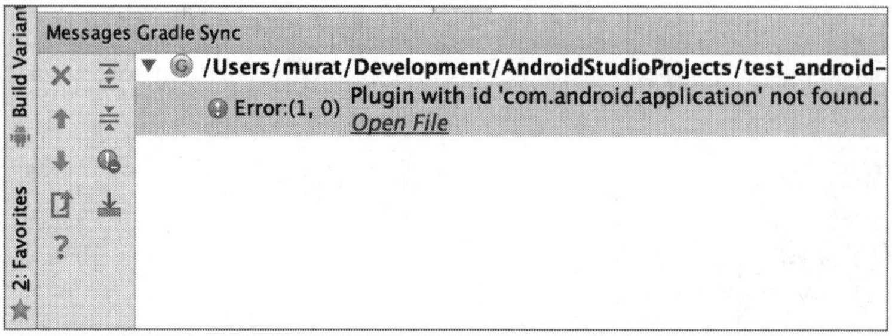

FIGURE 11-14: Gradle build failure after changing the plugin

Next, you can update module scope build.gradle file. There are some structural changes you need to implement, introduced by the experimental Gradle plugin. The most major change is the plugin name, which should be the first line in the gradle file. Change `apply plugin: com.android.application` to `apply plugin: 'com.android.model.application'`. Another major change is the `model` block, which wraps the `android` block. As you remember from the sample application, you are trying to change your imported project to be usable with the experimental plugin.

```
model {
    android {
        ...
    }
}
```

The experimental Gradle plugin also introduces some syntactic changes on configurations inside the Android block. You would need to change the proguard configuration to the following:

```
proguardFiles.add(file("proguard-rules.pro"))
```

The full code for the build.gradle module appears in Listing 11-4.

LISTING 11-4: Experimental plugin changes

```
apply plugin: 'com.android.model.application'

model {
    android {
        compileSdkVersion 23
        buildToolsVersion "23.0.2"

        defaultConfig {
            applicationId "com.expertandroid.ndkapplication"
            minSdkVersion.apiLevel 15
            targetSdkVersion.apiLevel 23
            versionCode 1
            versionName "1.0"
        }
        buildTypes {
            release {
                minifyEnabled false
```

```
                    proguardFiles.add(file("proguard-rules.pro"))
                }
            }
        }
    }

    dependencies {
        compile fileTree(dir: 'libs', include: ['*.jar'])
    }
```

You have completed all the necessary modifications for the new Gradle plugin syntax. However, you are still missing an important artifact—your NDK module name. The NDK module is declared as shown in Listing 11-5.

LISTING 11-5: NDK module declaration

```
apply plugin: 'com.android.model.application'

model {
    android {
        compileSdkVersion 23
        buildToolsVersion "23.0.2"

        defaultConfig {
            applicationId "com.expertandroid.ndkapplication"
            minSdkVersion.apiLevel 15
            targetSdkVersion.apiLevel 23
            versionCode 1
            versionName "1.0"
        }
        buildTypes {
            release {
                minifyEnabled false
                proguardFiles.add(file("proguard-rules.pro"))
            }
        }

        ndk {
      moduleName "ndkModule"
  }
    }
}
```

The experimental Gradle plugin introduces some other syntax changes to Gradle DSL. Although we covered the most common changes, you may refer to `http://tools.android.com/tech-docs /new-build-system/gradle-experimental` for the full set of changes.

In order to see more complex NDK project import samples, refer to Chapter 13, where we work on the vendor-provided NDK samples and import them to Android Studio using the rules covered here. Some examples are more complex with external libraries and some are as easy as changing the Gradle file.

Building Android NDK Projects

Building and packaging Android projects is pretty straightforward and does not introduce any more complexity, provided that your project has already implemented the DSL differences for the experimental Gradle plugin. This section covers some of the Gradle configurations that you may need to configure your builds.

The first change, which we covered in the previous section, is for configuring ProGuard. Because ProGuard is needed both for obfuscation and shrinking your APK, this configuration should be taken as a mandatory one rather than optional.

The experimental Gradle plugin introduces some changes to DSL syntax, which is shown in bold in the following snippet.

```
proguardFiles.add(file("proguard-rules.pro"))
```

Another change we have already seen but not covered in detail is to create and declare `productFlavors`. Product flavors are an essential part of working with NDK because native code is sensitive to device architecture unlike VM managed Java code. Product flavors from our sample project that create different products for different platform architectures are listed in Listing 11-6.

LISTING 11-6: Android NDK product flavors

```
android.productFlavors {
    create("arm") {
        ndk.abiFilters.add("armeabi")
    }
    create("arm7") {
        ndk.abiFilters.add("armeabi-v7a")
    }
    create("arm8") {
        ndk.abiFilters.add("arm64-v8a")
    }
    create("x86") {
        ndk.abiFilters.add("x86")
    }
    create("x86-64") {
        ndk.abiFilters.add("x86_64")
    }
    create("mips") {
        ndk.abiFilters.add("mips")
    }
    create("mips-64") {
        ndk.abiFilters.add("mips64")
    }
    // To include all cpu architectures, leaves abiFilters empty
    create("all")
}
```

By default, Android Studio assumes C/C++ code is placed in the `src/main/java` directory. However, C/C++ source code can be customized from Gradle, as shown in Listing 11-7.

LISTING 11-7: Native C/C++ code directory declaration in Gradle

```
model {
    android {
        ...
        sources {
            main {
                jni {
                    source {
                        srcDir "customSrc"
                    }
                }
            }
        }
    }
}
```

Most tool and compiler-related configuration is performed in the `model.android.ndk` block. The following items are some of the options you may choose to configure. Please note that these items are case sensitive and should be used as they appear.

➤ `moduleName`—Name of NDK module.

➤ `toolchain`—Toolchain used by NDK, llvm, or gcc. If you write `"clang"` on this parameter, the NDK build system will use the LLVM compiler.

➤ `toolChainVersion`—Version of the toolchain. There might be versions like 3.7, 2.8, and so on. You can change the version with this parameter.

➤ `CFlags.add("...")`—Environment variables needed by the C compiler.

➤ `cppFlags("...")`—Environment variables needed by the C++ compiler.

➤ `ldFlags("...")`—Library flags for the linker.

➤ `stl "..."`—Standard Template Library options.

Finally, you can set ABI specific configurations using a `model.android.abis` block. The following code snippet shows disabling SSSE3 instructions for x86 architecture.

```
android.abis {
    create("x86"){
        cppFlags.add("-DENABLE_SSSE3")
    }
}
```

Because the Gradle plugin for NDK integration is still experimental, changes to DSL syntax should be expected with new releases of the plugin.

ANDROID NDK PROJECTS RELEASE AND DEPLOYMENT

Android NDK projects used to rely on Android make files, and today, many Android projects with NDK modules still use `android.mk` files for the build process. However, Gradle offers an easier and single way to manage dependencies, automated tests, and the build/release cycle.

Although Android runs on different architectures, VM managed code, which is usually written in Java, is abstracted from the hosting platform. This gives the ability to deploy the same code to different architectures and delegates the interpretation problem to the VM. However, C/C++ code built with NDK is not managed by the VM and may require additional steps to preserve compatibility.

One main problem with NDK builds was integrating shared library (`.so`) files into your project. The new Android Studio and Gradle offers a more flexible way to handle `.so` files.

To include a `.so` file, create a folder named jniLibs under the `src/main` folder. Each target platform architecture is represented with a folder inside jniLibs, as shown in Listing 11-8.

LISTING 11-8: Library folders

```
- src/main
            - jniLibs
                      - amreabi
                   - mylib.so
                      - mips
                   - mylib.so
                      - x86
                   - mylib.so
```

The current jniLibs folder is the default location to place platform-dependent code. However, this location can be changed via Gradle. To declare a custom folder for `.so` files, add the following line to your build.gradle.

```
android {
    sourceSets.main {
        jniLibs.srcDir 'src/main/libs'
    }
}
```

This declaration will result in the following change to your folder structure:

```
- src/main
            - libs
                      - amreabi
                   - mylib.so
                      - mips
                   - mylib.so
                      - x86
                   - mylib.so
```

You finished adding your native libs and code to your project, but depending on the size of your native code, you may have introduced another problem to your project by creating a huge mono-lithic APK.

Multi vs. Fat Android Application APKs

Packaging all native code for each platform into one APK is not necessarily a bad thing and actu-ally might help keep your builds and versioning simple. However, if the native code and libraries included in your APK grow in size, the size of the APK grows with a multiplier of each platform, which may introduce unnecessary network traffic and disk usage.

By default, Gradle packages all native code and libraries into one *fat* APK. Basically, if you don't worry about the APK size, you may choose to continue with the defaults. However, if you want to split native code into platform-dependent APKs, you would need to add a product flavor for each APK, as shown in the following code. Version code has to be dynamically adapted when you have multiple APKs

```
def versionCodeBase = 11;
def versionCodePrefixes = ['armeabi': 1, 'armeabi-v7a': 2, 'arm64-v8a': 3,
 'mips: 5, 'mips-64': 6, 'x86: 8, 'x86-64': 9];

android.productFlavors {
    create("arm") {
        ndk.abiFilters.add("armeabi")
        versionCode = versionCodePrefixes.get("armeabi", 0) * 1000000 +
versionCodeBase
    }
    create("arm7") {
        ndk.abiFilters.add("armeabi-v7a")
        versionCode = versionCodePrefixes.get("armeabi- v7a", 0) * 1000000 +
versionCodeBase
    }
    create("arm8") {
        ndk.abiFilters.add("arm64-v8a")
        versionCode = versionCodePrefixes.get("arm64-v8a ", 0) * 1000000 +
versionCodeBase

    }
    create("x86") {
        ndk.abiFilters.add("x86")
        versionCode = versionCodePrefixes.get("x86", 0) * 1000000 + versionCodeBase
    }
    create("x86-64") {
        ndk.abiFilters.add("x86_64")
        versionCode = versionCodePrefixes.get("x86_64", 0) * 1000000 +
versionCodeBase
    }
    create("mips") {
        ndk.abiFilters.add("mips")
        versionCode = versionCodePrefixes.get("mips ", 0) * 1000000 +
versionCodeBase
    }
    create("mips-64") {
```

```
                ndk.abiFilters.add("mips64")
                versionCode = versionCodePrefixes.get("mips64", 0) * 1000000 +
        versionCodeBase
            }
```

Now you can choose any product flavor to build, run, or package platform-specific APKs. If you still need a fat APK among platform-specific ones, add the following product flavor to include all native code in one APK.

```
android.productFlavors {
    create("arm") {
        ndk.abiFilters.add("armeabi")
    }
    create("arm7") {
        ndk.abiFilters.add("armeabi-v7a")
    }
    create("arm8") {
        ndk.abiFilters.add("arm64-v8a")
    }
    create("x86") {
        ndk.abiFilters.add("x86")
    }
    create("x86-64") {
        ndk.abiFilters.add("x86_64")
    }
    create("mips") {
        ndk.abiFilters.add("mips")
    }
    create("mips-64") {
        ndk.abiFilters.add("mips64")
    }
    // To include all cpu architectures, leaves abiFilters empty
    create("fat")
        }
```

Fat APKs can be useful for development and CI builds where platform APKs would help you to have a smaller footprint in terms of network and storage of your app.

SUMMARY

In this chapter we covered Android NDK, which is an essential tool to unleash the performance and graphic capabilities of Android. Native code may be needed for deeper hardware integration, so there is no guarantee that an Android developer would never need to learn the basics of NDK. You have seen how Android Studio offers NDK integration via the experimental Gradle plugin. We focused on configuration and differences in Gradle DSL. We also covered how to integrate a piece of C/C++ code with your project.

Finally, we focused on different packaging options for projects that consist of native code and libraries. You learned how to separate platform-dependent code to minimize network and storage usage as well as how to package the app into a fat APK.

12

Writing Your Own Plugin

WHAT'S IN THIS CHAPTER?

➤   What are plugins?

➤   IntelliJ plugin architecture

➤   Writing your own plugin

➤   Interacting with UI

➤   Integrating with editor

➤   Packing and distributing plugins

WROX.COM CODE DOWNLOADS FOR THIS CHAPTER

The wrox.com code download for this chapter is found at www.wrox.com/go/expertandroid on the Download Code tab. The code for this chapter is in the Chapter12.zip file.

Because Android Studio is great as is, you might think you would never need to implement a new feature. But you might need to write a plugin to meet a custom need and unleash your full development potential.

This chapter focuses on the architecture of IntelliJ plugins and shows you how to write a plugin that can act without code, resources, a UI, or even a build cycle. The ability to write your own plugins will enable you to fix missing functionality. You can even share your plugins with other developers.

WAR STORY: MANY WAYS TO SKIN A CAT

Murat was the team lead in a banking project that had multi-language support. The localization files were simple Java property files, each containing a locale. One supported locale was Turkish, which is known for some nonstandard characters such as ğ, ş, and ı. Those characters needed to be represented in Unicode to display correctly. When he joined the project, he was surprised to see that everyone had their own method of converting Turkish characters to Unicode. One team member wrote his own desktop app, another used a website, and someone else printed out the Unicode values and entered them manually. In addition to the Unicode problem, the language files were large and many items were duplicated.

Murat decided to write his own plugin to add Eclipse-based vendor-specific tools. The requirements were simple: The plugin needed to listen to keystrokes, convert Turkish characters into Unicode values, and order them by value so that contributors could see whether a character was already available so duplicates would be avoided. After reviewing Eclipse plugin documents, he wrote his first Eclipse plugin, which did what was needed perfectly. The plugin seamlessly integrated with the locale file editor and did its job in each entry. Very soon his little plugin became a standard tool in every development environment. He decided to simplify the code and wrote a blog post about it, which soon became popular, at `https://dzone .com/articles/real-world-eclipse-plugins-two`.

INTELLIJ IDEA PLUGIN ARCHITECTURE

All popular IDEs—Eclipse, NetBeans, and, of course, IntelliJ Idea—support adding functional extensions and integrating them easily with the platform. Those functional extensions, called plugins, add new functionality and provide their own extensions for future plugins to introduce new functionality. The IntelliJ platform acts like a giant Lego platform in which other Lego blocks can be added by plugging into the right extensions.

All IDEs face the same problems. They need to work on different OSes, which introduce different file systems, while abstracting projects, runtimes, and even compilers from each other. In addition, they suffer from a problem that all applications have: an unresponsive UI resulting from time-consuming computations in the UI thread. This problem can become extreme because compiling and packaging code can be quite time-consuming when compared to other tasks performed by applications.

IDEs offer their own APIs and approaches to solve those issues. All IDEs, including IntelliJ, introduce their own project/file system, controlling time-consuming tasks, communicating with the UI when needed, and finding solutions to similar problems.

Actions

Actions are the most basic building blocks for interacting with users. An *action* is a selectable item that informs the platform that the user wants to trigger something. Creating new projects or opening existing ones are actions baked into the IDE. IntelliJ offers actions that can be added easily. Actions can be added though the UI or by adding a declaration to the `plugin.xml` file. `ActionEvent`, which carries data about the action itself (such as a selected file or text), is passed to the `actionPerformed` method related to the action.

In this section, you create a simple action. You start by downloading and installing IntelliJ, then move to adding an action with a declaration in the `plugin.xml` file, and finally you learn how to add an action using the wizards.

Creating a Simple IntelliJ Plugin

To start developing plugins for IntelliJ, first you need to download IntelliJ. Both the Community and Ultimate editions of IntelliJ Idea are capable of building plugins; however, we demonstrate the process with the Community edition because it is free of charge. From the JetBrains website at `https://www.jetbrains.com/`, click the IntelliJ IDEA link. From the page that opens, click the download link. Then click the Community Download option, as shown in Figure 12-1.

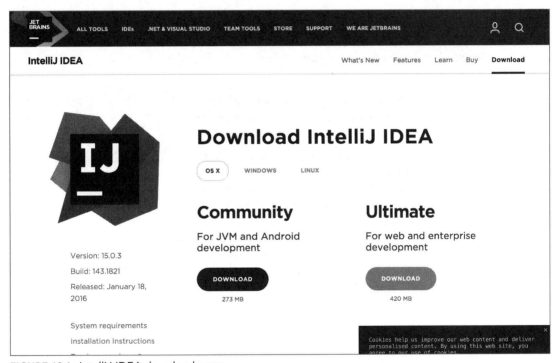

FIGURE 12-1: IntelliJ IDEA download page

Installing IntelliJ Idea is identical to installing Android Studio. (Refer to Chapter 1 for information on how to complete the installation once the download is complete.)

After IntelliJ is installed, follow these steps to add a plugin.

1. Click the IntelliJ Idea icon to start the IDE. You should already be familiar with IntelliJ Idea because it is the basis of Android Studio.

2. Select Create New Project, as shown in Figure 12-2.

FIGURE 12-2: IntelliJ IDE Start screen

3. In the window that opens, select the IntelliJ Platform Plugin option from the list in the left pane, as shown in Figure 12-3.

4. Because building plugins for IntelliJ means you need to build an IntelliJ within IntelliJ, point to the IntelliJ Idea folder as the IntelliJ Platform SDK location, as shown in Figure 12-4. This folder can be the same IntelliJ instance you have been working with as well as a newer or older version of IntelliJ to target another version of the SDK platform. This gives developers the freedom to develop the next version of an IDE inside the IDE.

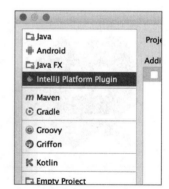

FIGURE 12-3: List of project types

Once the SDK is configured, the Plugin option will appear for running the project, as shown in Figure 12-5.

5. Click Run. IntelliJ Idea starts a new instance and displays the Welcome to IntelliJ IDEA wizard shown in Figure 12-6. You may close the new instance for now because it does not yet offer any new functionality.

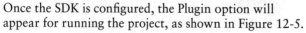

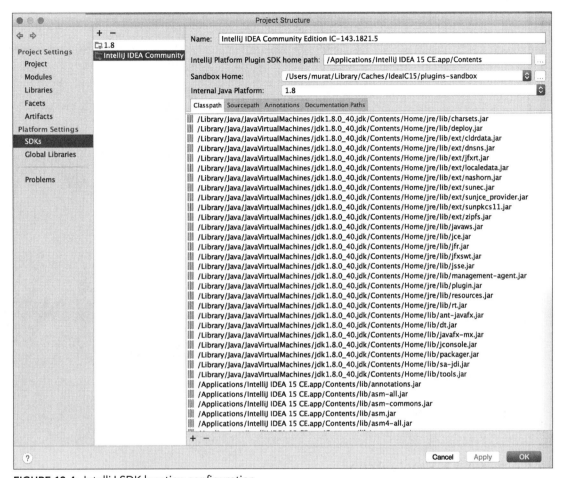

FIGURE 12-4: IntelliJ SDK location configuration

FIGURE 12-5: Plugin option on the IntelliJ toolbar

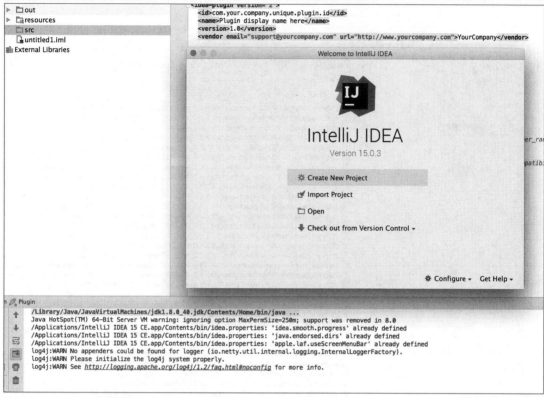

FIGURE 12-6: IntelliJ IDEA welcome wizard

Implementing a Simple Action

You have just built an IntelliJ Idea inside IntelliJ Idea, but because you haven't yet implemented anything, there is no new functionality added to the base platform of the current SDK. So let's implement the new, simple action.

1. Expand the resources folder, open `META-INF`, and select `plugin.xml`, as shown in Figure 12-7.

2. Because you want to add a new action, you need to declare your action inside `<action>` tags as seen in Listing 12-1. Type the following code to declare a simple action.

LISTING 12-1: Action declaration

```
<actions>
        <!-- Add your actions here -->
        <group id="MyPlugin.ExtendedMenu" text="_Extended Menu"
```

```
description="Extended menu">
                <add-to-group group-id="MainMenu" anchor="last" />
                <action id="Myplugin.ProjectInfo"
class="com.expertandroid.plugin.InfoPopup" text="Popup" description="project
info" />
        </group>
</actions>
```

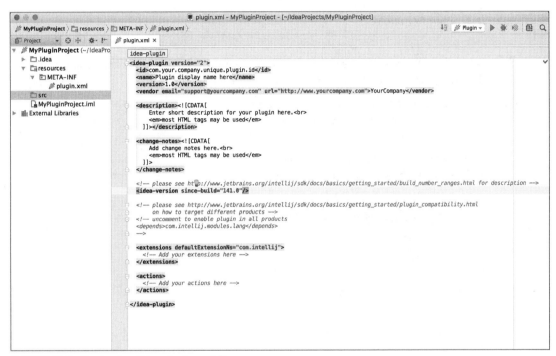

FIGURE 12-7: Contents of the plugin.xml file

Your declaration starts with a `<group>` tag, which can be used to group several actions. This is a good technique, one that is followed by many applications such as Word to group similar actions such as cut, copy, and paste into the Edit group, for example. Next, you use the `<add-to-group>` tag to declare where you want to hook your group. In this example, your new group will appear as the last item of the MainMenu.

Finally, it is time to declare the action itself. Your new action declares a unique ID, a class that is delegated to run when the action is clicked, text that will be shown in the menu, and finally the description of your new action.

The `plugin.xml` editor is smart enough to analyze and validate the XML file. You may have noticed that all fields are marked green except for the class field of the action, which is normal because you don't have the class yet. When focused on the class name, IntelliJ even offers a smart fix, as shown in Figure 12-8.

FIGURE 12-8: Smart fix options

3. Select the first option to create an action class with the name and package you already declared in the XML. Android Studio will create a new class in the correct package.

4. Navigate to the new InfoPopup class and double-click to open. In order to respond as an action, the InfoPopup class must extend the AnAction class.

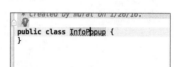

5. Check whether InfoPopup extends the AnAction class and, if not, as shown in Figure 12-9, use the smart assistance provided by the editor, as shown in Figure 12-10.

FIGURE 12-9: Class inheritance control

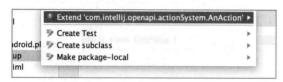

FIGURE 12-10: Smart assistance to help class extension

Every class that extends AnAction class needs to implement an actionPerformed method. This method received an AnActionEvent parameter, which carries all the data related to the action. Listing 12-2 shows a simple example to get the project object data from the event and finally display a message dialog box greeting us with the project name.

LISTING 12-2: Example class to get object data from events

```
package com.expertandroid.plugin;

import com.intellij.openapi.actionSystem.AnAction;
import com.intellij.openapi.actionSystem.AnActionEvent;
import com.intellij.openapi.actionSystem.PlatformDataKeys;
import com.intellij.openapi.project.Project;
import com.intellij.openapi.ui.Messages;

public class InfoPopup extends AnAction {
    public void actionPerformed(AnActionEvent event) {
```

```
        Project project = event.getData(PlatformDataKeys.PROJECT);
        Messages.showMessageDialog(project, "Hi, welcome to "+project.getName(),
    "Project Info", Messages.getInformationIcon());
        }
}
```

6. Now, it is time to try your plugin. Click Run and start a new IntelliJ instance that consists of your new plugin. Choose New Project and start the IDE. You are free to choose any project name but keep in mind that the name you have chosen for the project will be displayed by your action. Check the top menu bar for the Extended Menu menu, which should be the last item in the list, as shown in Figure 12-11.

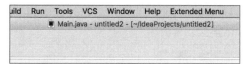

FIGURE 12-11: New menu view

7. Click the Extended Menu to display its item. The list should display only one item, shown in Figure 12-12, which you added to the XML file.

8. Select Popup from the drop-down menu. This action should trigger your action class, which retrieves project metadata and displays a popup to greet you with the project name, as shown in Figure 12-13.

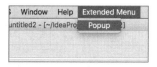

FIGURE 12-12: New menu item

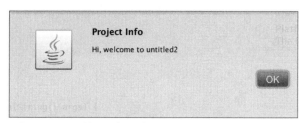

FIGURE 12-13: New plugin's popup greeting

That's it! You have completed your first plugin and customized the IntelliJ. Your plugin doesn't perform much yet, but this is a gateway to many capabilities to extend the IDE.

Alternatively, you can use the New Action wizard to create your plugin rather than editing the XML manually. Right-click anywhere on the project and select New ⇨ Action, as shown in Figure 12-14.

This option will bring up the New Action dialog box, which offers a GUI editor to tweak all the available settings related to the action you want to create. You may add a new action and choose the Extended Menu group you previously created to add a new action, as shown in Figure 12-15.

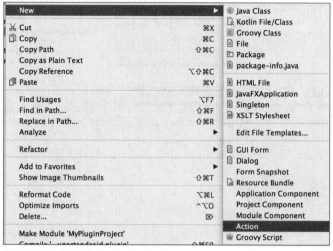

FIGURE 12-14: New action selection

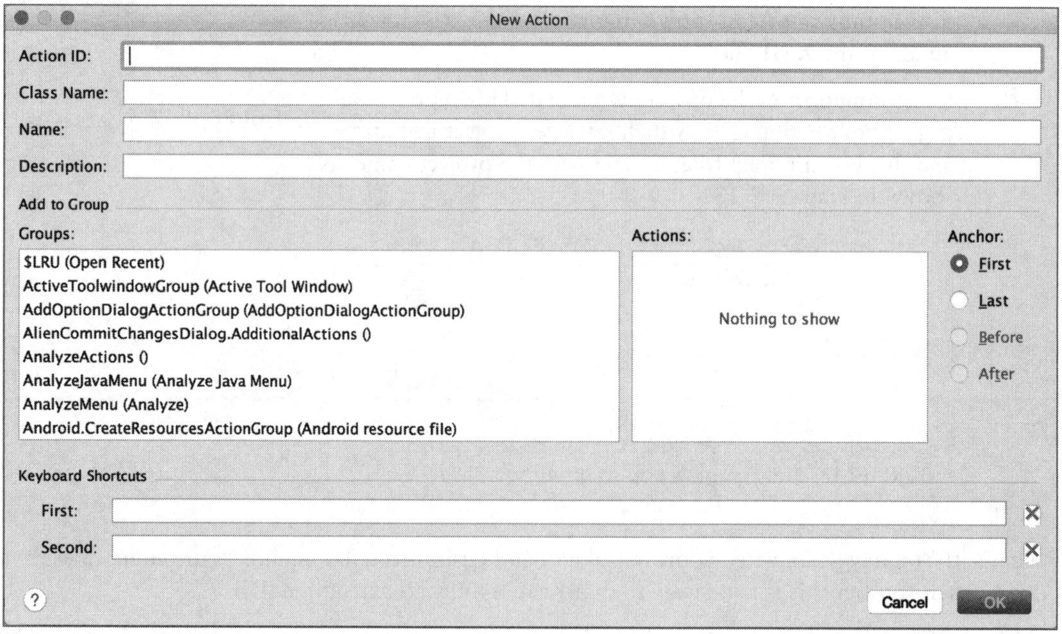

FIGURE 12-15: New Action window

The end product will be no different when using the wizard than when adding a declaration to the `plugins.xml` file.

Threading

IntelliJ does not enforce strict control over what you can't do on the UI thread, but this does not mean you should be adventurous. Window managers run a single thread to interact with users, including collecting the input and presenting the output. As a general rule, to keep an application responsive, you as a developer should not lock the UI thread by performing lengthy operations. Ideally anything that is not directly interacting with a user through the UI shouldn't be performed in the UI thread.

Plugins should execute tasks by passing a runnable by calling the `executeOnPooledThread` method of `ApplicationManager`, as shown in the following code.

```java
ApplicationManager.getApplication().executeOnPooledThread(new Runnable() {
    @Override
    public void run() {
        //...
    }
});
```

The runnable this provides will be executed in the background without blocking the UI thread. But wait a minute—because this code will run in the background thread, how can you return and perform something on the UI thread? The `executeOnPooledThread` method returns a `Future<T>` reference object, which can be used for asynchronous response. However, constantly checking the `Future` object would also lock the UI thread and create a similar problem.

A proper way to return to the UI thread is with an API similar to `executeOnPooledThread`. Passing a runnable object to the `invokeLater` method of `ApplicationManager`, as shown in the following code, will delegate the execution of the runnable to UI thread.

```java
ApplicationManager.getApplication().invokeLater(new Runnable() {
    @Override
    public void run() {
        //...
    }
});
```

As shown in the following code, IntelliJ also offers `runReadAction` and `runWriteAction` methods in `ApplicationManager` that similarly take a runnable and execute read/write operations without blocking the UI thread.

```java
ApplicationManager.getApplication().runReadAction(new Runnable() {
    @Override
    public void run() {
        //...
    }
});
ApplicationManager.getApplication().runWriteAction (new Runnable() {
    @Override
    public void run() {
        //...
    }
});
```

File System

Abstracting the project and file structure is crucial to handling different operating systems and file systems. The IntelliJ platform introduces several concepts to handle file-related operations.

➤ **Virtual files**—IntelliJ offers VFS (Virtual File System) for representing files on a file system in the com.intellij.openapi.vfs package. Most file operations such as read/write are done via the `VirtualFile` class. VFS reference of a file can be gathered from actions, directly from paths or documents, and from PSI files, which we cover last in this list. Although read/write operations are the most well-known interactions with files, listening for changes or updates about the file becomes more important when IDEs are involved. VFS also offers a very simple mechanism to be notified about changes in the file system by adding a `VirtualFileListener` via `addVirtualFileListener`. VirtualFiles represent already existing files in the file system, so they cannot be created programmatically.

➤ **Documents**—Documents represent the contents of VirtualFiles. The contents of the Document are editable. Document references can be accessed from actions, VirtualFiles, and PSI files. Documents are weak objects that are dynamically created when contents of a VirtualFile is accessed, but unlike VirtualFiles, they are eligible for garbage collection if not referenced. Unlike VFS, new Documents can be created programmatically via `EditorFactory.createDocument()`, which will create a new PSI. Changes in the contents of Documents can be listened for with a `DocumentListener` by calling `addDocumentListener`. Global listeners, which observe changes on all documents, can also be added over the `EditorFactory.getEventMulticaster().addDocumentListener` method. Document contents can be modified with several different methods, such as `setText`, `insertString`, and `replaceString`.

➤ **Program Structure Interface (PSI) files**—PSI files represent files but with hierarchical elements based on programming language syntax. There are specific PSI implementations such as PsiJavaFile, XmlFile, PyFile, and more. Unlike VFS and Documents, which are application scoped, PSI files are project scoped. This way, each project can work on its own PSI instance for a file, which is shared among projects. PSIs consist of PSI elements, which form a tree structure to represent the contents with respect to the programming language.

A PSI document reference can be accessed via a VirtualFile reference, Document reference, an action, or by a child element belonging to the target PSI. PSI documents can also be searched in a project's `getFilesByName` method from the `FileNameIndex` class by passing project reference, search name, and the scope. PSI files can be created using `PSIFactory`. Changes can be listened for by adding a `PsiTreeChangeListener`. Thanks to the hierarchical structure, it is easy to navigate between PSI elements and perform modification on elements instead of files. Elements can be iterated using a `PsiRecursiveElementWalkingVisitor` object.

Projects and Components

Now that you know about IntelliJ plugin architecture, you can write a more complex plugin. If you have read the war story sidebar earlier in this chapter, you know about the special Turkish characters that are not compatible with anything but Unicode. Although this example seems to target a very specific task that covers a small percentage of the whole world's app ecosystem, it clearly demonstrates how to write a plugin that listens for changes in a file, responds to them, and reads or writes file contents.

Let's start by designing a plugin and separating tasks. First, you need to listen for changes in files. You already know that IntelliJ Idea has its own file system API, so you can expect to find something that suits your needs in the API.

The first step in building a new plugin is to declare it in the `plugin.xml`, as you did in the "Implementing a Simple Action" section earlier in this chapter. In that example, you built an action. This time, your plugin needs to do more than just receive actions. This plugin needs to integrate into the IDE and check project files for changes. IntelliJ Idea's plugin architecture offers project components that suit this purpose. Each project component implements the `ProjectComponent` interface, which introduces the following project and component lifecycle methods as well as the project instance.

➤ `void initComponent()`—Entry point for component related initialization code.

➤ `void disposeComponent()`—Exit point for component related disposal code.

➤ `String getComponentName()`—Returns component name.

➤ `void projectOpened()`—Project lifecycle method that is called when the project is opened.

➤ `void projectClosed()`—Project lifecycle method that is called when the project is closed.

➤ `ProjectComponent (Project project)`—Constructor for concrete `ProjectComponent` implementation. Reference to the project object is passed at the time of initialization.

To listen for file changes, you need to start, register, and unregister your plugin. The `initComponent` and `disposeComponent` methods are the perfect candidates for that purpose. Next, you need to listen for file changes. You need to target project files, which exist in the OS's file system, and because you don't need to create and work with new files, Virtual File System looks suitable for the job. VFS offers the `VirtualFileListener` interface, which can be registered to listen to a variety of file-related events.

➤ `void propertyChanged(VirtualFilePropertyEvent)`—Called when properties of the file have changed.

➤ `void beforePropertyChange(VirtualFilePropertyEvent)`—Called before the property change action takes place. Think of this method as a hook that is executed just before the change.

➤ `void contentsChanged(VirtualFileEvent)`—Called when contents of the file have changed. Typically occurs when the file is accessed via an editor.

➤ `void beforeContentsChange(VirtualFileEvent)`—Called before the contents change action takes place. Think of this method as a hook that is executed just before the contents change.

➤ `void fileCreated(VirtualFileEvent)`—Called when a new file is created. Typically occurs when a wizard or another plugin creates or generates a new file.

➤ void fileDeleted(VirtualFileEvent)—Called when an existing file is deleted. Typically occurs when the user or another plugin deletes an existing file.

➤ void beforeFileDeletion (VirtualFileEvent)—Called before the file deletion action takes place. Think of this method as a hook that is executed just before the deletion.

➤ void fileMoved(VirtualFileEvent)—Called when the location of an existing file has been changed by moving the file. Typically occurs as a result of refactoring or dragging a file to another package or folder.

➤ void beforeFileMovement (VirtualFileEvent)—Called before the file move action takes place. Think of this method as a hook that is executed just before the file is moved.

➤ void fileCopied(VirtualFileEvent)—Called when a file is copied to another location while maintaining the original copy in the original location. Typically occurs when the user copies/pastes a file.

Although there are many different event methods, it is clear that you need to be targeting the contentsChanged method to listen for changes to a file and take action. The action should be simple enough to search for specific characters and replace them with specific codes, and because you know when the contents of the file change, you can easily hook your functionality there. Let's start by writing a simple method to go over a string and replace each "special" character with the specified Unicode value. Listing 12-3 shows a simple but naïve implementation to do the job. (Note that this string conversion can be done more efficiently with regular expressions, but that is beyond the scope of this chapter.)

LISTING 12-3: Unicode replace with special characters

```java
private String convertTr(String nativeText) {
    Map charMap = new HashMap<>();
    charMap.put("s", "\\\u00e7");
    charMap.put("ç", "\\\u00e7");
    charMap.put("Ç", "\\\u00c7");
    charMap.put("ğ", "\\\u011f");
    charMap.put("Ğ", "\\\u011e");
    charMap.put("ş", "\\\u015f");
    charMap.put("Ş", "\\\u015e");
    charMap.put("ı", "\\\u0131");
    charMap.put("İ", "\\\u0130");
    charMap.put("ö", "\\\u00f6");
    charMap.put("Ö", "\\\u00d6");
    charMap.put("ü", "\\\u00fc");
    charMap.put("Ü", "\\\u00dc");
    String unicodeTxt = "";
    if (nativeText != null) {
        unicodeTxt = new String(nativeText);
        Set keySet = charMap.keySet();
        Iterator it = keySet.iterator();
        while (it.hasNext()) {
            String nativeChar = (String) it.next();
```

```
                        String unicodeChar = (String) charMap.get(nativeChar);
                        unicodeTxt = unicodeTxt.replaceAll(nativeChar, unicodeChar);
                }
        }
        return unicodeTxt;
    }
```

Now you can implement the `VirtualFileListener` to extract the string contents from the file when a change occurs. As we discussed earlier, the target method is `contentsChanged`. However, because the `VirtualFileListener` interface offers a long list of methods, you need to create the method bodies even if you don't plan to do anything when the related event occurs, as shown in Listing 12-4.

LISTING 12-4: VirtualFileListener implementation

```
VirtualFileListener listener= new VirtualFileListener() {
        @Override
        public void propertyChanged(@NotNull VirtualFilePropertyEvent
virtualFilePropertyEvent) {

        }

        @Override
        public void contentsChanged(@NotNull VirtualFileEvent virtualFileEvent) {


                try {
InputStream is = virtualFileEvent.getFile().getInputStream();
                        BufferedReader reader = new BufferedReader(new
InputStreamReader(is));
                        String line = null;
                        String finalLine = "";
                        while ((line = reader.readLine()) != null) {
                            //concanate the lines
                            finalLine += line + "\n";
                        }
                        //we are done with the input stream
                        is.close();
                        //convert chars
                        String toOut = convertTr(finalLine);
                        if (!toOut.equals(finalLine)) {
                          virtualFileEvent.getFile().setBinaryContent(toOut.getBytes());
                        }
                } catch (IOException e) {
                    e.printStackTrace();
                } finally {

                }

        }
```

```java
        @Override
        public void fileCreated(@NotNull VirtualFileEvent virtualFileEvent) {

        }

        @Override
        public void fileDeleted(@NotNull VirtualFileEvent virtualFileEvent) {

        }

        @Override
        public void fileMoved(@NotNull VirtualFileMoveEvent virtualFileMoveEvent) {

        }

        @Override
        public void fileCopied(@NotNull VirtualFileCopyEvent virtualFileCopyEvent) {

        }

        @Override
        public void beforePropertyChange(@NotNull VirtualFilePropertyEvent
    virtualFilePropertyEvent) {

        }

        @Override
        public void beforeContentsChange(@NotNull VirtualFileEvent
    virtualFileEvent) {

        }

        @Override
        public void beforeFileDeletion(@NotNull VirtualFileEvent virtualFileEvent) {

        }

        @Override
        public void beforeFileMovement(@NotNull VirtualFileMoveEvent
    virtualFileMoveEvent) {

        }
    };
```

Each time a change occurs, your contentsChanged implementation will be called. It reads the file content using the FileChangeEvent and calls your function that converts the characters in the stream. Finally, you write the file contents back to the file. Although you implemented all the logic behind your plugin, you haven't yet hooked your VirtualFileListener to your component. To achieve that, you need to register the event listener when the component is initialized and unregister it when the component is disposed of, as shown in Listing 12-5.

LISTING 12-5: Register and unregister the VirtualFileListener

```java
import com.intellij.openapi.components.ProjectComponent;
import com.intellij.openapi.project.Project;
import com.intellij.openapi.vfs.*;
import org.jetbrains.annotations.NotNull;

import java.io.*;
import java.util.HashMap;
import java.util.Iterator;
import java.util.Map;
import java.util.Set;

public class MyProjectComponent implements ProjectComponent {
    public MyProjectComponent(Project project) {
    }

    @Override
    public void initComponent() {
        // TODO: insert component initialization logic here

        VirtualFileManager.getInstance().addVirtualFileListener(listener);
    }

    @Override
    public void disposeComponent() {
        // TODO: insert component disposal logic here
        VirtualFileManager.getInstance().removeVirtualFileListener(listener);
    }

    @Override
    @NotNull
    public String getComponentName() {
        return "MyProjectComponent";
    }

    @Override
    public void projectOpened() {
        // called when project is opened
    }

    @Override
    public void projectClosed() {
        // called when project is being closed
    }
}
```

The plugin code is ready, but you haven't yet added the description to plugins.xml. Component declaration is much simpler and more straightforward than an action. Open plugin.xml and add the declaration in Listing 12-6.

LISTING 12-6: Plugin declaration

```
<project-components>
  <component>
    <implementation-class>com.expertandroid.plugin.
MyProjectComponent</implementation-class>
  </component>
</project-components>
```

Now create the class `MyProjectComponent` inside the com.expertandroid.plugin package and paste the code you have written so far. Your plugin is ready to be used. But wait a minute—previously we said that read and write operations should be done through `Runnable` with the provided API so as not to block the UI thread. So let's refactor the code to make use of `runReadAction` on reads and `runWriteAction` on writes. Listing 12-7 has the complete code listing we have covered in pieces so far.

LISTING 12-7: Read and write action refactoring

```java
package com.expertandroid.plugin;

import com.intellij.openapi.application.ApplicationManager;
import com.intellij.openapi.components.ProjectComponent;
import com.intellij.openapi.project.Project;
import com.intellij.openapi.vfs.*;
import org.jetbrains.annotations.NotNull;

import java.io.*;
import java.util.HashMap;
import java.util.Iterator;
import java.util.Map;
import java.util.Set;

public class MyProjectComponent implements ProjectComponent {
    public MyProjectComponent(Project project) {
    }

    @Override
    public void initComponent() {
        // TODO: insert component initialization logic here

        VirtualFileManager.getInstance().addVirtualFileListener(listener);
    }

    @Override
    public void disposeComponent() {
        // TODO: insert component disposal logic here
        VirtualFileManager.getInstance().removeVirtualFileListener(listener);
    }

    @Override
    @NotNull
```

```java
    public String getComponentName() {
        return "MyProjectComponent";
    }

    @Override
    public void projectOpened() {
        // called when project is opened
    }

    @Override
    public void projectClosed() {
        // called when project is being closed
    }

    VirtualFileListener listener= new VirtualFileListener() {
        @Override
        public void propertyChanged(@NotNull VirtualFilePropertyEvent
virtualFilePropertyEvent) {

        }

        @Override
        public void contentsChanged(@NotNull VirtualFileEvent virtualFileEvent) {

                ApplicationManager.getApplication().runReadAction(new Runnable() {
                    @Override
                    public void run() {
                        try {
                            InputStream is =
virtualFileEvent.getFile().getInputStream();
                            BufferedReader reader = new BufferedReader(new
InputStreamReader(is));
                            String line = null;
                            String finalLine = "";
                            while ((line = reader.readLine()) != null) {
                                //concanate the lines
                                    finalLine += line + "\n";
                            }
                            //we are done with the input stream
                              is.close();
                            //convert chars
                              String toOut = convertTr(finalLine);
                            if (!toOut.equals(finalLine)) {
                                write(virtualFileEvent, toOut);
                            }
                        } catch (IOException e) {
                            e.printStackTrace();
                        } finally {

                        }
                    }
                });

        }
```

```java
        private void write(VirtualFileEvent virtualFileEvent, String toOut){
            ApplicationManager.getApplication().runWriteAction (new Runnable() {
                @Override
                public void run() {
                    try {
                        virtualFileEvent.getFile().setBinaryContent(toOut.
getBytes());
                    } catch (IOException e) {
                        e.printStackTrace();
                    } finally {

                    }
                }
            });
        }

        private String convertTr(String nativeText) {
            Map charMap = new HashMap<>();
            charMap.put("s", "\\\\u00e7");
            charMap.put("ç", "\\\\u00e7");
            charMap.put("Ç", "\\\\u00c7");
            charMap.put("ğ", "\\\\u011f");
            charMap.put("Ğ", "\\\\u011e");
            charMap.put("ş", "\\\\u015f");
            charMap.put("Ş", "\\\\u015e");
            charMap.put("ı", "\\\\u0131");
            charMap.put("İ", "\\\\u0130");
            charMap.put("ö", "\\\\u00f6");
            charMap.put("Ö", "\\\\u00d6");
            charMap.put("ü", "\\\\u00fc");
            charMap.put("Ü", "\\\\u00dc");
            String asciiText = "";
            if (nativeText != null) {
                asciiText = new String(nativeText);
                Set keySet = charMap.keySet();
                Iterator it = keySet.iterator();
                while (it.hasNext()) {
                    String nativeChar = (String) it.next();
                    String asciiChar = (String) charMap.get(nativeChar);
                    asciiText = asciiText.replaceAll(nativeChar, asciiChar);
                }
            }
            return asciiText;
        }

        @Override
        public void fileCreated(@NotNull VirtualFileEvent virtualFileEvent) {

        }

        @Override
        public void fileDeleted(@NotNull VirtualFileEvent virtualFileEvent) {
```

```
        }

        @Override
        public void fileMoved(@NotNull VirtualFileMoveEvent virtualFileMoveEvent) {

        }

        @Override
        public void fileCopied(@NotNull VirtualFileCopyEvent virtualFileCopyEvent) {

        }

        @Override
        public void beforePropertyChange(@NotNull VirtualFilePropertyEvent
virtualFilePropertyEvent) {

        }

        @Override
        public void beforeContentsChange(@NotNull VirtualFileEvent
virtualFileEvent) {

        }

        @Override
        public void beforeFileDeletion(@NotNull VirtualFileEvent virtualFileEvent) {

        }

        @Override
        public void beforeFileMovement(@NotNull VirtualFileMoveEvent
virtualFileMoveEvent) {

        }
    };

}
```

Now click Run to fire an IntelliJ with your new plugin to give a test drive. Create a new file or open an existing file and type one of the special characters that your plugin covers and watch how the file dynamically updates its contents.

That's it—you have written your first useful plugin. So far, you have seen how to define actions and how to respond to file events but you haven't actually yet interacted with the editor.

Editors

Your plugin currently reads the entire contents of the file and writes back the modified string. This strategy might suffer as the file size gets larger. Alternatively, you can hook an action into your editor to run your algorithm on a selection of text and replace it with the Unicode values. This way, you can have better control on what parts of the text are being changed and also run your algorithm in a more efficient way.

1. Create a new plugin project as you did in the previous example and choose New Window, as shown in Figure 12-16.

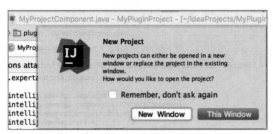

FIGURE 12-16: New plugin creation

2. Open the `plugin.xml` file and add the Action declaration using the `<action>` tag, as shown in Listing 12-8.

LISTING 12-8: Unicode plugin declaration

```
<action id="MyPlugin.EditorAction"
class="com.expertandroid.plugin.EditorAction" text="Replace"
        description="Replaces characters with unicode values">
  <add-to-group group-id="EditorPopupMenu" anchor="first"/>
</action>
```

You have just added a new action labeled `"Replace"` as the first item in the editor popup menu. Because you haven't created the `EditorAction` class yet, it is marked in red in Figure 12-17.

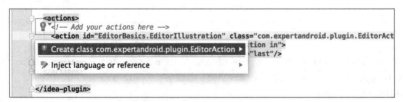

FIGURE 12-17: Creating the EditorAction class

3. Click the Create class option from the contextual help to create your action class. You will use the contextual help to implement the needed methods in your new class.

4. Select Implement methods, as shown in Figure 12-18.

5. Now modify your existing code from the previous example to work with the editor action.

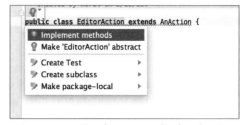

FIGURE 12-18: Implement methods selection

This time you will use the Document API instead of VFS. Documents are structural files so you can easily use the selection made in the editor. To access the selected text, you need the active instance of the Project, Editor, and the Document. You will use the Document for accessing the file contents and the editor to access the `SelectionModel` object, which will let us interact with the editor and the selection.

```
final Editor editor = anActionEvent.getRequiredData(CommonDataKeys.EDITOR);
final SelectionModel selectionModel = editor.getSelectionModel();
final int start = selectionModel.getSelectionStart();
final int end = selectionModel.getSelectionEnd();
```

In Listing 12-9, you access the Editor from the action event and get the beginning and ending indexes of the selection via the `SelectionModel` object of the editor. Now you can get the document instance and change the selection by using your previous `convertUnicode` method. Listing 12-9 shows the complete code of the `actionPerformed` method.

LISTING 12-9: actionPerformed method implementation

```
@Override
public void actionPerformed(AnActionEvent anActionEvent) {
    final Editor editor = anActionEvent.getRequiredData(CommonDataKeys.EDITOR);
    final SelectionModel selectionModel = editor.getSelectionModel();
    final int start = selectionModel.getSelectionStart();
    final int end = selectionModel.getSelectionEnd();

    final Project project = anActionEvent.getRequiredData(CommonDataKeys.PROJECT);
    final Document document = editor.getDocument();

    WriteCommandAction.runWriteCommandAction(project, new Runnable() {
        @Override
        public void run() {
            String unicodeText=convertUnicode(selectionModel.getSelectedText());
            document.replaceString(start, end, unicodeText);
        }
    });
    selectionModel.removeSelection();
}
```

You may have noticed that you used `runWriteCommandAction` from the `WriteCommandAction` class instead of `runWriteAction` from `ApplicationManager`. Both methods execute write actions in a separate thread, but this time you are executing write from an action instead of an application component. Now you can copy the `convertUnicode` method from the previous example to complete the missing piece.

Everything looks ready, but what if the selection is empty? Your selection model may not even exist because the user may not be working with an editor all the time. You need to make your plugin safe by checking if the editor is open and has a valid selection. To achieve this goal, you need to implement an `update` method in your `Action` class. Listing 12-10 gets the current instance of the project and the editor and enables the presentation of your action only if there is a selection made in the editor.

LISTING 12-10: Update action

```
@Override
public void update(AnActionEvent anActionEvent){
    final Project project = anActionEvent.getData(CommonDataKeys.PROJECT);
    final Editor editor = anActionEvent.getData(CommonDataKeys.EDITOR);

    if (project != null && editor != null){
        anActionEvent.getPresentation().setVisible(editor.getSelectionModel().
hasSelection());
    }else{
        anActionEvent.getPresentation().setVisible(false);
    }
}
```

Finally, you are ready to test your plugin. Click Run and start a new project or open an existing one. Open the editor and right-click after making a text selection, as shown in Figure 12-19.

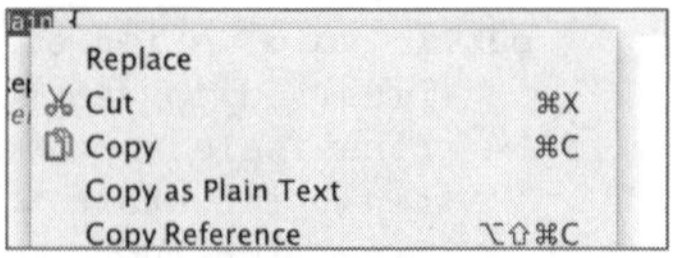

Once you click Replace, your plugin will kick in and replace the special characters in the text selection with given Unicode values. If no selection is made, the Replace action will not appear on the action menu, as shown in Figure 12-20.

FIGURE 12-19: The plugin action with text selected

You may use the same strategy to easily generate or edit code, either by listening for changes in files or waiting for user action.

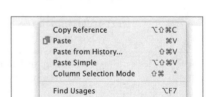

FIGURE 12-20: Right-click actions when no text is selected

Wizards

Wizards are a well-known part of the user experience when software needs to complete a task that consists of one or more steps. IntelliJ Idea offers an easy way to create wizards for your custom tasks.

  1.   Declare an extension by adding the extension in Listing 12-11 to your `plugin.xml`.

LISTING 12-11: EXTENSION DECLARATION

```
<extensions defaultExtensionNs="com.intellij">
    <!-- Add your extensions here -->
    <moduleBuilder builderClass="com.expertandroid.CustomWizard"
id="Custom.Wizard" order="first"/>
</extensions>
```

2. Use the contextual help to create a `CustomWizard` class, as shown in Figure 12-21.

FIGURE 12-21: CustomWizard class creation

3. Find and open the newly created `CustomWizard` class.

Wizards in IntelliJ Idea extend from the `ModuleBuilder` class. Because your class is also extending the same base class, you need to implement missing methods.

4. Use the contextual help as shown in Figure 12-22 to create the two missing methods in Figure 12-23.

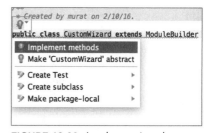

FIGURE 12-22: Implementing the CustomWizard action

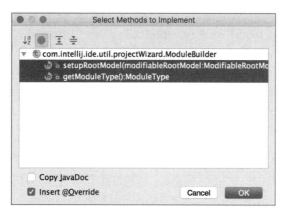

FIGURE 12-23: ActionWizard class methods

The first method, `setupRootModel`, is where project-specific setup such as setting the compiler, libraries, and default folder and files is done. Let's leave this module as is because building a complex custom project is beyond the scope of this chapter. However, if you are interested in learning more about what can be done on this method of the JavaModuleBuilder class, which is responsible for creating Java projects from IntelliJ source code, checking the source can give you a better understanding (https://upsource.jetbrains.com/idea-ce/file/idea-ce-1731d054af4ca27aa827c03929e27eeb0e6a8366/java/openapi/src/com/intellij/ide/util/projectWizard/JavaModuleBuilder.java).

The second method you implemented is getModuleType, which returns the type of the Module. There are predefined ModuleTypes in IntelliJ, such as StdModuleTypes.JAVA for common project types.

5. For now, change the getModuleType method to return ModuleType.EMPTY, as shown in Listing 12-12.

LISTING 12-12: Set ModuleType return value

```java
public class CustomWizard extends ModuleBuilder {
    @Override
    public void setupRootModel(ModifiableRootModel modifiableRootModel)
throws ConfigurationException {

    }

    @Override
    public ModuleType getModuleType(){
        return ModuleType.EMPTY;
    }
}
```

6. Add some content to your wizard by adding steps.

Each step is defined with a ModuleWizardStep object, which will introduce its own UI elements and will update the data model with user-specified values. Listing 12-13 creates two ModuleWizardSteps.

LISTING 12-13: Creating ModuleWizardSteps

```java
ModuleWizardStep[] steps = new ModuleWizardStep[]{new ModuleWizardStep() {
    @Override
    public JComponent getComponent() {
        return new JFileChooser();
    }

    @Override
    public void updateDataModel() {

    }
}, new ModuleWizardStep() {
    @Override
    public JComponent getComponent() {
        return new JRadioButton("Enabled?");
```

```
        }

        @Override
        public void updateDataModel() {

        }
    }};
```

The first `ModuleWizardStep` returns a `JFileChooser` in the `getComponent` method. This method is responsible for building the UI for the wizard step. Because the return type is `JComponent`, you can easily create swing composites, which are basic Java UI elements for Java-based desktop applications. As we mentioned before, `updateDataModel` is responsible for reflecting the changes introduced by the user to the data model of the wizard. For the sake of keeping the example simple, you will leave it empty.

The second `ModuleWizardStep` returns a `JRadioButton`, which can also be used to modify the data model of the wizard.

7. Now add your wizard steps to your wizard.

The `ModuleBuilder` base class has a method named `createWizardSteps` that you need to override to return your new steps, as shown in Listing 12-14.

LISTING 12-14: CustomWizard step creation

```
@Override
public ModuleWizardStep[] createWizardSteps(@NotNull WizardContext
wizardContext, @NotNull ModulesProvider modulesProvider) {
    return steps;
}
```

The complete code for your custom wizard is given in Listing 12-15.

LISTING 12-15: CustomWizard complete implementation

```
package com.expertandroid;

import com.intellij.ide.util.projectWizard.ModuleBuilder;
import com.intellij.ide.util.projectWizard.ModuleWizardStep;
import com.intellij.ide.util.projectWizard.WizardContext;
import com.intellij.openapi.module.ModuleType;
import com.intellij.openapi.options.ConfigurationException;
import com.intellij.openapi.roots.ModifiableRootModel;
import com.intellij.openapi.roots.ui.configuration.ModulesProvider;
import org.jetbrains.annotations.NotNull;

import javax.swing.*;

/**
 * Created by murat on 2/10/16.
 */
public class CustomWizard extends ModuleBuilder {
```

```java
    @Override
    public void setupRootModel(ModifiableRootModel modifiableRootModel)
throws ConfigurationException {

    }

    @Override
    public ModuleType getModuleType() {
        return ModuleType.EMPTY;
    }

    ModuleWizardStep[] steps = new ModuleWizardStep[]{new ModuleWizardStep() {
        @Override
        public JComponent getComponent() {
            return new JFileChooser();
        }

        @Override
        public void updateDataModel() {

        }
    }, new ModuleWizardStep() {
        @Override
        public JComponent getComponent() {
            return new JRadioButton("Enabled?");
        }

        @Override
        public void updateDataModel() {

        }
    }};

    @Override
    public ModuleWizardStep[] createWizardSteps(@NotNull WizardContext
wizardContext, @NotNull ModulesProvider modulesProvider) {
        return steps;
    }
}
```

Now it is time to run and test your new wizard. Click Run plugin and select New Project from the welcome screen. Select Empty Project from the bottom of the list on the left pane, as shown in Figure 12-24.

Clicking Next will display the custom wizard steps, as shown in Figure 12-25.

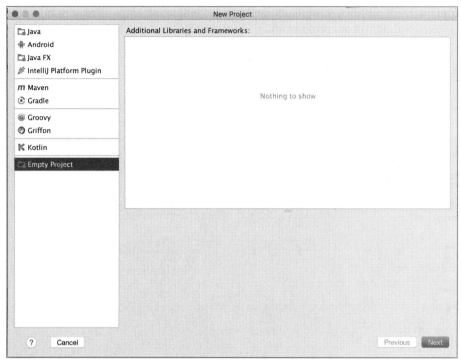

FIGURE 12-24: Creating the new wizard

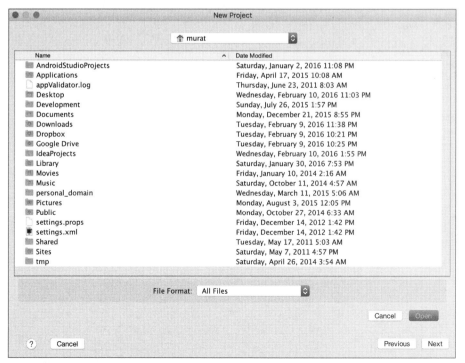

FIGURE 12-25: Custom wizard steps

Although creating wizards may not seem like something you will use daily, wizards can be very helpful when you're creating custom projects and custom settings.

PACKAGING AND DISTRIBUTION

Packaging and releasing your plugin is very easy and straightforward. IntelliJ packages plugins as JAR files that can be installed from external repositories or from the local disk. To package your plugin, right-click the project pane and select the Prepare Plugin Module option with your project name, as shown in Figure 12-26.

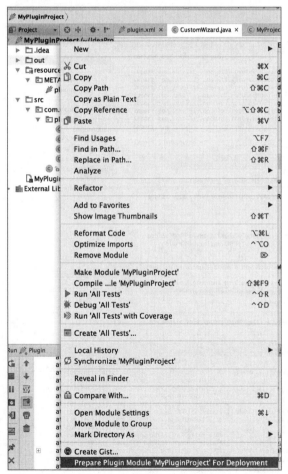

FIGURE 12-26: Plugin deployment selection

IntelliJ will compile, build, and package your plugin as a jar file and will display a popup dialog box stating it is ready for deployment, as shown in Figure 12-27. That's it, you have just finished building your plugin and you can upload your plugin into a repository or just make it available publicly as a jar file.

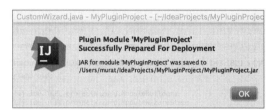

FIGURE 12-27: Plugin package info popup dialog box

Now let's install the plugin to Android Studio and give a test run. First locate the newly packaged jar file. You may refer to the URI, which was provided in the popup in Figure 12-27.

Open Android Studio and select Preferences. Highlight the Plugins option in the left pane in the Preferences window. As shown in Figure 12-28, click the Install Plugin from disk option at the bottom of the screen and locate the jar file.

FIGURE 12-28: Installing the plugin from disk

That's it. You have just installed your new plugin to Android Studio. Ideally, your plugin should be uploaded into a repository, which can be installed via the Browse repositories option.

SUMMARY

In this chapter, you learned how to build plugins for the IntelliJ Idea platform to extend Android Studio for your special needs. IntelliJ Idea offers an easy-to-extend architecture with many APIs to help you write your own plugin. The chapter started with building basic actions; then you learned how to listen to files and projects. You integrated a plugin with the editor to interact with its contents. We covered how to create custom wizards for specific tasks and how to customize their steps.

Finally, you learned how to package, load, and release your plugins for other developers' use.

13

Third-Party Tools

WHAT'S IN THIS CHAPTER?

➤   Android Studio plugins

➤   Intel System Studio

➤   Intel Integrated Native Developer Experience (INDE)

➤   Qualcomm Android software tools

➤   NVIDIA Android Software Tools

Throughout this book, we showcased all the fundamental features you would expect to use for Android application development and to expand the capabilities of Android Studio. In this chapter, we present some popular Android Studio plugins and chip vendor tools you can use to expand your development skills and capabilities.

We start by exploring Android Studio plugins and their use cases, and then we follow up with tools developed by Intel to enhance Android application development together with Android Studio. Finally, we look into Qualcomm and NVIDIA software tools for Android.

ANDROID STUDIO PLUGINS

Android Studio is extensible with plugins that can improve its capabilities and functionality with custom actions. It is possible to install plugins from plugin repositories online or via a local jar or zip file, as you did in Chapter 12.

You can navigate between plugins from the Android Studio start or Preferences windows. Open the Preferences window from Android Studio and navigate to the Plugins section to see the installed plugins, as shown in Figure 13-1.

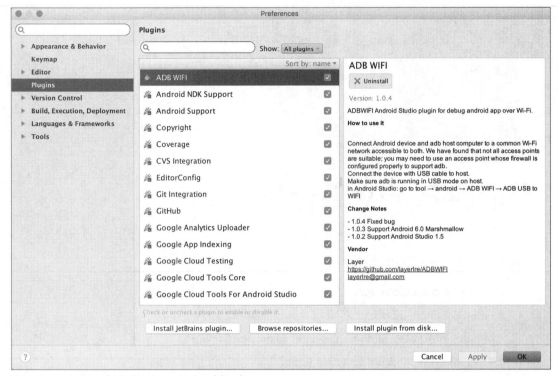

FIGURE 13-1: Installed plugins in Android Studio

This initial window shows just the installed plugins. By default, there are many plugins installed related to core Android application development and some other supporting tools to improve the development cycle, such as GitHub integration and Google Cloud Tools.

You can look at the available plugins by clicking the Browse repositories button. The window that opens, shown in Figure 13-2, shows the plugins from the default JetBrains repository.

You can also look at the JetBrains plugin website at `https://plugins.jetbrains.com/?androidstudio` to see the popular plugins, as shown in Figure 13-3.

The JetBrains plugins site makes it easier to find popular plugins because it breaks them out into categories. However, if you want to install them, you should use the Android Studio plugins window.

Installing a new plugin is easy; when you select the plugin from the list of plugins and click the green Install button, shown in Figure 13-2, Android Studio will download and install the plugin seamlessly. After the plugin has been installed, you will see the Restart Android Studio button, shown in Figure 13-4. Restarting is required to make Android Studio ready to use the installed plugin. After you restart Android Studio, the plugin will be activated. In this example, we installed the .ignore plugin.

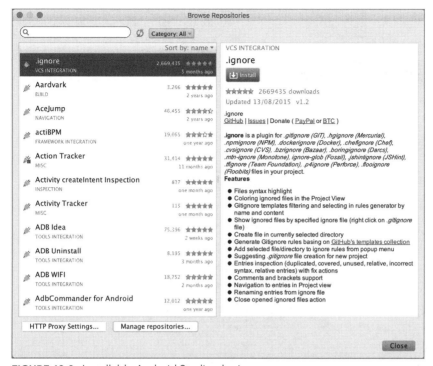

FIGURE 13-2: Installable Android Studio plugins

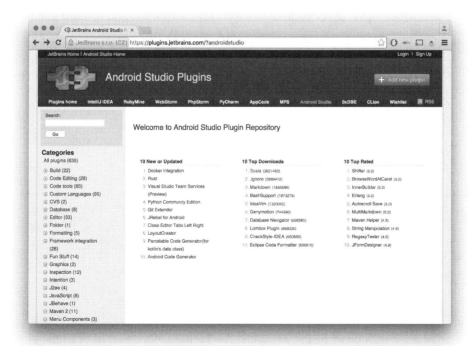

FIGURE 13-3: JetBrains plugins website

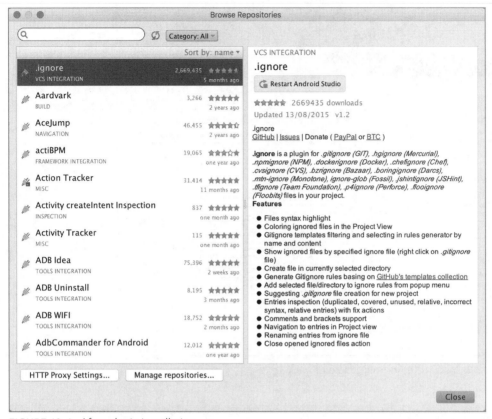

FIGURE 13-4: After plugin installation

The .ignore plugin's detailed description is shown in the right pane. .ignore is used to organize files that are autogenerated by the build system and add them to the `.ignore` file (`.gitignore` if you are using Git) to prevent adding them to the version control system.

As Figure 13-5 shows, the New menu now has an .ignore item that includes options for adding .gitignore or any other version control system's ignore file.

Here are some other plugins you may find useful for application development:

➤ **ADB Idea**—This tool enables Android Studio to perform certain ADB commands through Android Studio. When you install ADB Idea, it will be enabled in the Tools ⇨ Android menu, with the actions shown in Figure 13-6. ADB Idea makes it easy to control the debugging process and remote devices without launching the adb tool from the terminal.

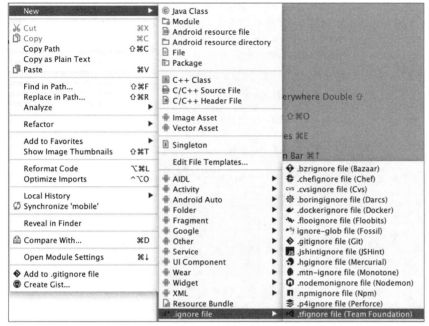

FIGURE 13-5: .ignore plugin

➤ **ADB WIFI**—This plugin helps you debug Android devices remotely over Wi-Fi. To enable your device for debugging over Wi-Fi, make sure your device is connected through USB first, and then navigate to Tools ➪ Android ➪ ADB WIFI and click ADB USB to WIFI, as shown in Figure 13-7.

When your device is ready, it will display the connection info with your device's local IP address, as shown in Figure 13-8. Both your development machine and Android device should be on the same network.

Avoiding the messiness of cables is relaxing, so the capability to debug your device wirelessly is a great feature. While it's easy to debug smartphones with a USB cable, wirelessly debugging Android TV is far better.

➤ **Android Drawable Importer**—Managing drawable resources can be tricky when your application needs multiple resolutions and colors. Android Drawable Importer

FIGURE 13-6: ADB Idea action list

FIGURE 13-7: ADB WIFI menu

FIGURE 13-8: ADB WIFI connection information

makes it easy to import icons with multiple colors, assets with multiple resolutions simultaneously, and so on. New actions (shown in Figure 13-9) are available from the New option, accessed by right-clicking, or from the File menu.

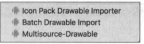

FIGURE 13-9: Android Drawable Importer plugin actions

As an example of how these options work, let's look at Icon Pack Drawable Importer. When you click this action, a window opens where you can select and customize icons for your application. You can set the color, name, and resolution of the icon, as shown in Figure 13-10.

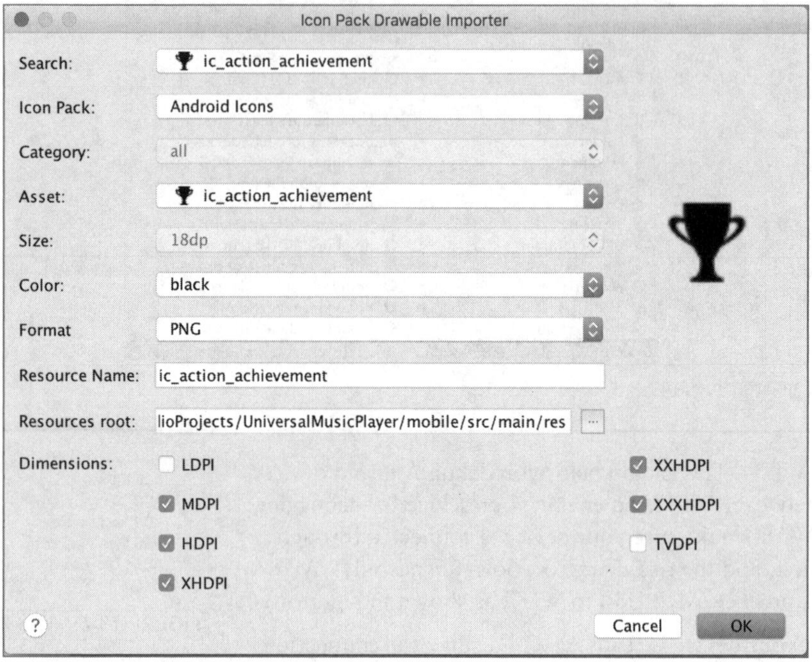

FIGURE 13-10: Icon Pack Drawable Importer

➤ **Android strings.xml tools**—If your applications have too many string resources for localizations, it might be hard to find one or to add a missing translation string. This basic tool does two simple actions to sort the strings and add the missing string resource for your localization files.

➤ **Android Parcelable Code Generator**—In Android, serialization to pass data between objects is made with the Parcelable interface, so you should implement the Parcelable interface for the class you want to serialize. Implementation of the Parcelable class has a pattern to follow, and the Android Parcelable Code Generator helps you create the methods and fields to implement it. When this plugin is installed, you can generate Parcelable code within a Java class by selecting Generate ➪ Parcelable from the Generate menu, as shown in Figure 13-11.

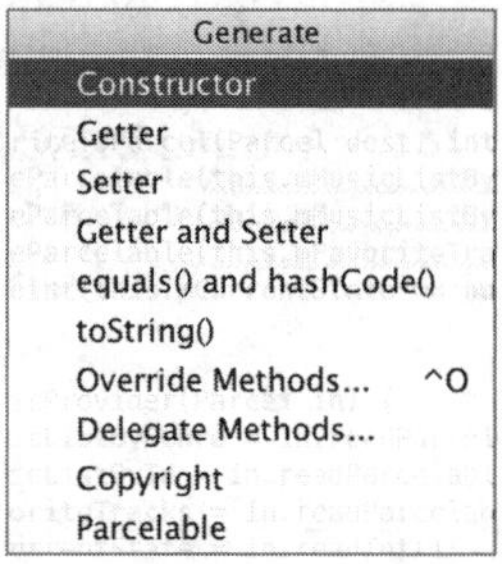

FIGURE 13-11: Generate menu for the Parcelable class

> **NOTE** *Refer to the Android API if you are new to Parcelable or serialization concepts:* `http://developer.android.com/reference/android/os/Parcelable.html`.

➤ **Android Holo Colors**—With Android Holo Colors, you can easily create custom XML resources with a desired color to use in layouts. After Android Holo Colors is installed, a new "H" button is enabled on the toolbar, as shown in Figure 13-12. When you click on the H button, the window shown in Figure 13-12 opens so you can configure your new XML resource with the selected Holo color.

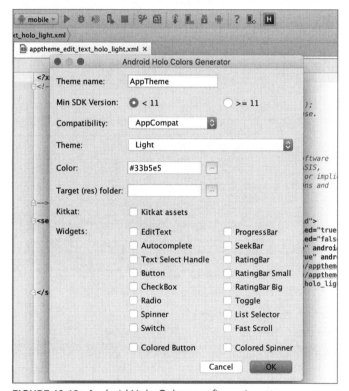

FIGURE 13-12: Android Holo Colors configuration page

When the configuration is ready, select the destination resource folder, which is your application's project root folder. According to your configuration, an XML file will be created under the drawable folders such as `apptheme_btn_check_holo_dark.xml`.

This plugin also has an online version, which can be found at `http://android-holo-colors.com`. On this website, you can generate the required XML resources and download them as a zip file to add to your project.

➤ **Key Prompter**—You might find this tool to be annoying, but it is useful to learn the keyboard shortcuts. It works by displaying a popup like
the one shown in Figure 13-13 when you click an action on an Android
Studio menu that has a shortcut key.

FIGURE 13-13: Key
Prompter popup

➤ **jimu Mirror**—This plugin allows you to dynamically design and develop
user interfaces for the Android UI. While you are editing XML layouts,
you can see previews instantly. jimu Mirror gives you the ability to immediately see changes
on the emulator and the device.

Right after you install the jimu Mirror plugin and restart Android Studio, you can either
start a 30-day free trial or buy it from `http://jimulabs.com`.

After installation, jimu Mirror installs a new menu and toolbar buttons, as shown in
Figure 13-14.

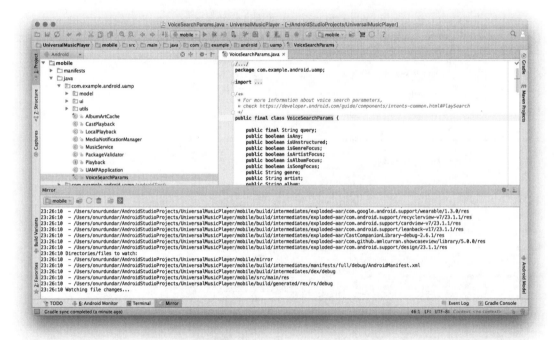

FIGURE 13-14: jimu Mirror tool menu and action buttons

You can start and stop Mirror from the toolbar or the tool window with the Start button.
The first time you start jimu Mirror, it will both install and start the jimu Mirror app on
your device. When it starts, you will see actions in the jimu Mirror window and the files
that are being generated and sent to the Android device. After all files are generated and

ready, jimu Mirror will list the layouts on the device so you can easily work on them and immediately see any changes on the device.

➤ **Genymotion**—Genymotion is a third-party Android emulator to manage testing Android applications on virtual devices with the provided Java API. Developers and teams can choose to use Genymotion instead of AVD. Genymotion's advantage is that it uses a Java API to test your application on a virtual copy of commercial Android devices from vendors such as Samsung and HTC.

Genymotion provides multiple pricing and licensing options; you can see them all at `https://www.genymotion.com/pricing-and-licensing/`. If you just want to test Genymotion, the Basic version is free for personal use. To download Genymotion, you first need to create an account on the site.

After Genymotion is installed, set the Genymotion path to the Android Studio plugin from the Preferences window, under Other Settings ⇨ Genymotion.

Now, when you click the Genymotion button on the toolbar, the Genymotion Device Manager window opens. This list will be empty initially, but after you create virtual devices in the Genymotion application, they will be listed as shown in Figure 13-15.

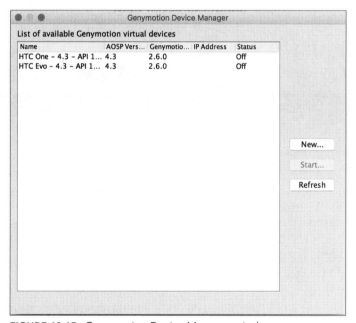

FIGURE 13-15: Genymotion Device Manager window

To create your first virtual device, click the New button to start the Genymotion application. In the window that opens, you select the new virtual device, as shown in Figure 13-16.

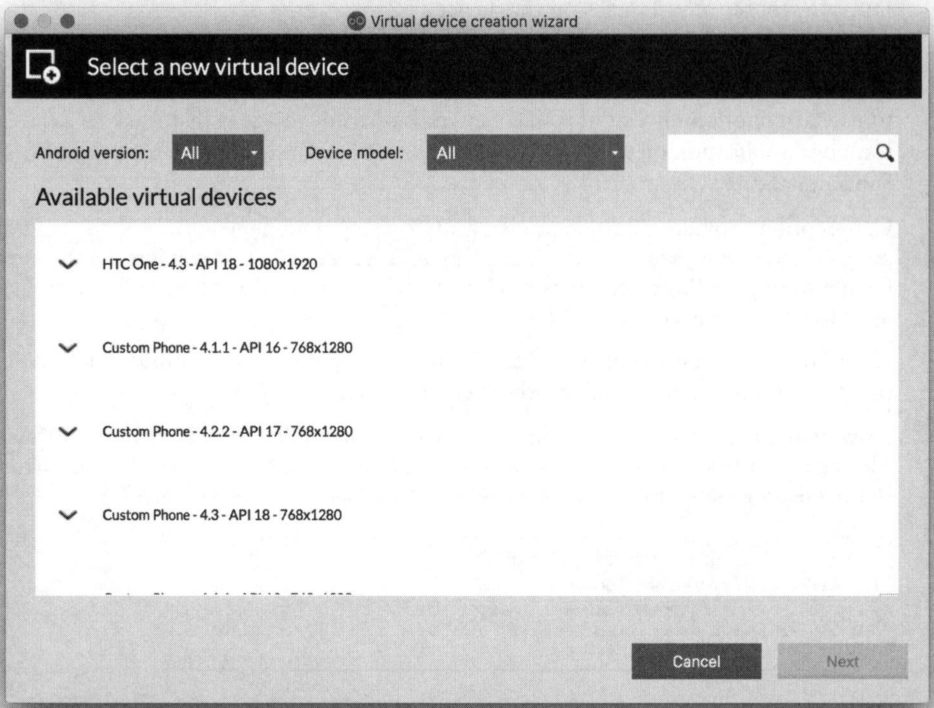

FIGURE 13-16: Genymotion Virtual Device List

You can get the real devices' exact images as virtual devices with Genymotion. AVD provides only the core Android SDK with the Google API, but because most popular Android phone producers—Samsung, HTC, Asus, Motorola—have a customized Android OS on the device, you can get the exact device's virtual image only with Genymotion.

When you finish creating the selected Android virtual devices, you will see them in the Genymotion Device Manager window as well as inside the Genymotion application itself, as shown in Figure 13-17.

When you run the selected virtual device, you will see it in the Genymotion Player window, as shown in Figure 13-18. Actual performance is really good on Genymotion. You can also see tools in the player to adjust GPS, use the camera, and so on in the Genymotion Player window.

FIGURE 13-17: Genymotion application window

FIGURE 13-18: Genymotion Player window

There is one more thing to mention: You can use Genymotion's Java API to test instrumentation. You can find the Java API Guide at `https://docs.genymotion.com/pdf/PDF_Java_API/Java-API-1.0.2-Guide.pdf`.

Detailed documentation for the Genymotion Java API can be found at `https://cloud.genymotion.com/static/external/javadoc/index.html`. The Java API helps with instrument testing of the battery, GPS, SMS, and phone calls on the device.

INTEL'S ANDROID SOFTWARE TOOLS

As one of the chip vendors providing a hardware platform to develop Android devices, Intel provides tools and software libraries to help device manufacturers and developers. These tools help engineers tune all aspects of Android performance, from Android device drivers to services and applications.

Intel System Studio

Intel System Studio is a pack of software tools provided for engineers who are going to develop Intel architecture based devices. Software tools from System Studio can be used to fine tune applications and the Android OS itself. Native applications can be developed with Intel C compiler. Tools also include the performance libraries MKL, IPP, and TBB, and analyzer applications for deep performance analysis of graphics and threads.

Intel System Studio's main target is Android device manufacturers, system integrators, and embedded application developers; however, the tools and libraries (see Figure 13-19) can also be used by application developers to improve performance, especially applications with native code.

Category	Component	Linux*1,2 Composer Edition	Professional Edition	Ultimate Edition	Android*2 Composer Edition	Professional Edition	Ultimate Edition	Windows* Composer Edition	Professional Edition	Ultimate Edition	VxWorks*3 VxWorks Edition	FreeBSD* FreeBSD Edition
Host Operating Systems		Linux* Windows*			Linux* Windows*			Windows*			Linux* Windows*	Linux* FreeBSD*
Integrated Development Environment		Eclipse*, WR Workbench*			Eclipse*			Visual Studio*			WR Workbench*	
Compiler & Libraries	Intel® C++ Compiler	○	○	○	○	○	○	○	○	○	○	○
	Intel® Integrated Performance Primitives	○	○	○	○	○	○	○	○	○	○	
	Intel® Math Kernel Library	○	○	○				○	○	○		
	Intel® Threading Building Blocks	○	○	○	○	○	○	○	○	○		
System & Application Debuggers	Intel® System Debugger			○			○				○7	
	Intel® Debug Extensions for WinDbg*4									○		
	Intel®-enhanced GDB® Application Debugger	○	○	○	○	○	○					
	Intel® Debugger for Heterogeneous Compute	○	○	○	○	○	○					
Performance, Power & Correctness Analyzers	Intel® VTune™ Amplifier⁵		○	○		○	○		○	○		○
	Intel® Energy Profiler		○	○		○	○		○	○		
	Intel® Inspector		○	○					○	○		
	Intel® Graphics Performance Analyzers		○	○		○	○		○	○		

1 Linux*, Embedded Linux*, Wind River* Linux*, Yocto Project*
2 Linux* and Android* target support available in a single product
3 Available from Wind River* with Wind River* VxWorks* platform
4 Via Intel* ITP-XDP3 probe, OpenOCD*, Intel* SVT Closed Chassis Adapter* and EDKII* for UEFI*
5 Also available for OS X* host as a separate download
7 Intel* System Debugger provides VxWorks* OS awareness – available with Ultimate Editions

FIGURE 13-19: Intel System Studio version comparison

Intel System Studio is available on Linux and Windows hosts. From `https://software.intel.com/en-us/intel-system-studio`, select Android as the target OS and pick the version of Intel System Studio (Composer, Professional, or Ultimate) to continue the download. During installation, when you are asked for a serial number you must register to be able to download and start a 30-day evaluation period. That's covered later in this section.

You can check the Product Brief to see detailed descriptions of tools provided with Intel System Studio at `https://software.intel.com/sites/default/files/managed/18/d8/intel-system-studio-2016-product-brief_final.pdf`.

The features of each version of Intel System Studio are available on the website (refer to Figure 13-19).

For this example, we downloaded the Ultimate 2016 version for Linux: `system_studio_2016.1.030.tar`. Let's extract the file and start GUI-based installation.

```
$ tar xvf  system_studio_2016.1.030.tar
$ cd system_studio_2016.1.030
```

To install the required tools, the fxload and gcc-multilib packages should be installed. Our host machine is running Ubuntu 14.04, so we will use apt-get. If the packages are not installed, installation will warn you about the missing dependencies:

```
$ sudo apt-get install fxload gcc-multilib
```

Start GUI-based installation from the system_studio_2016.1.030 folder:

```
$ ./install_GUI
```

The installer asks about the rights of the tools for root or the current user. We selected the sudo installer to be on the safe side and installed the system tool for all users in the system. If you select sudo-based or root installation, you will be asked for the sudo password in the next window.

The list of tools will be shown in the Welcome window, as shown in Figure 13-20.

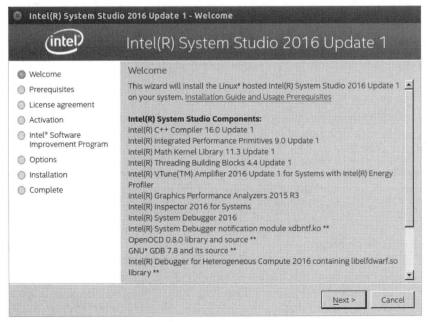

FIGURE 13-20: Intel System Studio components

If you already have a serial number, you can enter it in the next window, or just select Evaluate this product. The next screen shows the Installation Summary and is where you configure the installation directory. The default is s/opt/intel. The opt directory is used by third-party applications in Linux-based operating systems.

The next window, shown in Figure 13-21, presents the Eclipse IDE integration options because most Intel System Studio tools are integrated with Eclipse IDE. If you select integration with an existing Eclipse installation, Eclipse will be configured for use with Intel System Studio. We selected not to integrate with Eclipse, so we will see Eclipse Luna in the /opt/intel/eclipse directory.

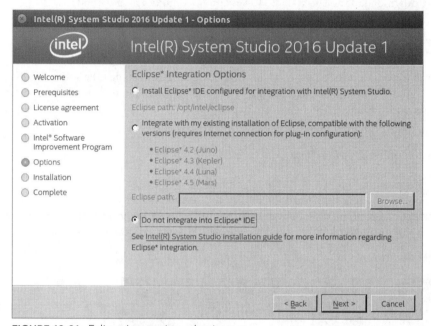

FIGURE 13-21: Eclipse Integration selection

When the installation process asks about Wind River integration, skip it. Finally you see the Android NDK integration screen; we integrated Intel's compiler with our Android NDK. If you want to do so, just type the location of ndk-bundle's toolchain folder path, as shown in Figure 13-22, and you are done preparing the installation.

If you downloaded the offline version as we did, it won't take long for the installation to finish. The online version could take longer because it will download the required files during the installation (rather than downloading them with the offline installer), and so depends on the speed of your Internet connection.

The next section discusses some tools provided with Intel System Studio for Android development use. The Intel C++ Compiler is covered in detail first because it is the main component for optimizing binaries for Intel devices. Then we briefly discuss the Intel Integrated Performance Primitives library, Intel Thread Building Blocks library, Intel VTune Amplifier, Energy Profiler, and Graphics Performance Analyzer so you understand their capabilities and purpose.

Intel's Android tools are targeted primarily at embedded software developers and native game developers, not developers of basic Android SDK applications.

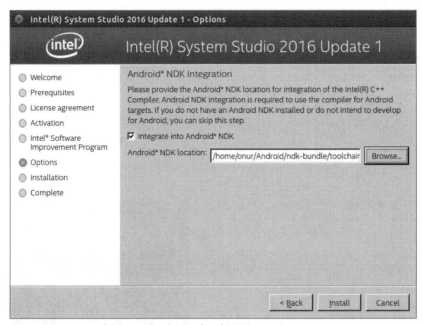

FIGURE 13-22: Intel System Studio Android NDK integration

Intel C++ Compiler

The Intel C++ Compiler (ICC) is included with Intel System Studio for use in Android. Intel's compiler generates only x86 and x86_64 native applications, so the Intel C++ compiler will generate only x86 and x86_64 binaries.

If you indicated the path correctly during installation, you will see x86-icc (write all folders) under the ndk-build/toolchains folder. However, you are not yet ready to use ICC to build x86 binaries; you need to complete the setup to ensure that switching from NDK's default x86 compiler to the Intel C++ Compiler happens.

> **NOTE** *If your NDK integration setup didn't work, you can navigate to the Intel System Studio installation directory and run the following command to complete installation:*
>
> ```
> $ ./opt/intel/ide_support/android_ndk/ndk_integration.sh /path/to/
> ndk-bundle
> ```
>
> Finally, you need to make sure compiler variables have been set up:
>
> ```
> $ source /opt/intel/compilers_and_libraries_2016.1.150/linux/bin/
> compilervars.sh -arch ia32 -platform android
> ```

Now ICC is ready to build C and C++ code. Android Studio is not ready to work with the Intel C++ Compiler, so for this example you manually build sample C code to see whether a test file you will load really compiled with Intel's compiler.

To begin, open the HelloJNI sample application from the NDK sample list (see Chapter 11 for information about native application samples). After you have loaded the HelloJNI sample, you will see the `hello-jni.c` file in the src/main/jni folder. We will add a new line to `hello-jni.c` to see whether it has been compiled with the Intel C++ compiler. Change the final return line as shown in the following code snippet:

```
#ifdef __INTEL_COMPILER_UPDATE
    return (*env)->NewStringUTF(env, "Hello from Intel C++ !");
#else
    return (*env)->NewStringUTF(env, "Hello from JNI !  Compiled with ABI "
ABI ".");
#endif
```

Open a terminal in Android Studio and navigate to the jni folder to create the `libhello-jni.so` library for x86:

```
$ cd  src/main/jni
$ icc –platform=android -c hello-jni.c
$ icc –platform=android -shared -o libhello-jni.
so hello-jni.o
```

Now copy the `.so` file to the jniLibs folder to make sure the shared library is copied to APK. Create a jniLibs folder under the src/main folder and an x86 folder under the jniLibs folder you just created. Then copy the shared library to jniLibs/x86.

```
$ cp src/main/jni/libhello-jni.so src/main/
jniLibs/x86
```

Now, you need to build the project and run the application on an x86 virtual device or a real device. This should show that it worked, as shown in Figure 13-23.

Using ICC manually as in this example is not very efficient. You can use it the "old fashioned" way (using Makefiles and `Android.mk` files) until there is support in Android Studio to select ICC for x86 in the gradle configuration. Until then, you need to use it this way or continue to use it in the Eclipse IDE.

ICC is also supported in Visual Studio, so you can build Android code with native libraries using Visual Studio and ICC.

FIGURE 13-23: Intel C++ Compiler Hello JNI sample

Intel Integrated Performance Primitives (Intel IPP)

Intel IPP for Android is available only with Intel System Studio; there is no standalone download for it.

The Intel IPP library is a very advanced set of code-based functions optimized for Intel's Streaming SIMD Extensions and Intel's Advanced Vector Extension instruction sets. IPP functions are the fundamental algorithms used in digital media, communications, and scientific, embedded software applications.

There are many areas in which Intel IPP can help you run complex algorithms and can help software run better on Android devices having Intel's SoC (System on Chip). The Intel IPP library uses an Intel CPU's advanced instruction sets to make software run faster and more energy efficiently.

Detailed documentation for Intel IPP can be found at `https://software.intel.com/en-us/intel-ipp`.

Intel System Studio installation places the Intel IPP libraries, headers, and samples in the /opt/intel/ipp folder.

Under the bin folder, you can find `ippvars.sh` to source the Intel IPP library and header files to the system. The bin folder also holds the examples folder for Android samples.

Intel ICC and IPP tools are not integrated with Android Studio yet so you should use Eclipse or traditional command-line build tools to create libraries with IPP.

IPP libraries provide the best functions to create good native applications, especially if you are going to work on audio processing, image recognition, or pattern recognition. IPP can help optimize applications that use low-level data processing from sensors such as microphones, cameras, and motion sensors.

Intel Thread Building Blocks (Intel TBB)

Intel TBB is a library to optimize C++ code for highly optimized parallelization. It is provided with Intel System Studio. Its files and examples are in the /opt/intel/tbb directory. Unlike Intel IPP, you can also download the open source version of the TBB library. The latest open source versions can be found at `https://www.threadingbuildingblocks.org/`.

Like ICC and IPP, TBB has a `tbbvars.sh` file for include and library directories.

TBB enhances parallelization of C++ code to optimize existing data structures and algorithms. You can find many samples inside the tbb/examples directory such as a concurrent hash map, priority queue, graph samples, parallel loop implementations, and a Sudoku solver. All are great examples to get you ready for parallel programming with TBB.

Intel VTune Amplifier

Intel System Studio delivers Intel's performance profiler to get detailed CPU and GPU data using Intel's VTune Amplifier. Intel VTune Amplifier is provided with Intel System Studio, and is also available as a standalone application. Visit `https://software.intel.com/en-us/intel-vtune-amplifier-xe` to download it for your target OS—currently Android is the only option—and use on a Linux or Windows host.

In our Intel System Studio installation, you should navigate to /opt/intel/vtune_amplifier_for_systems. There you will see bin32 and bin64 folders that include the binaries to run the standalone

Intel VTune Amplifier application. We will use the 64-bit version and run it with the following command:

```
$ ./bin64/amplxe-gui
```

The first time you run it, it will ask whether you want to participate in the developer program; you can either skip that or participate. Then you will see the window shown in Figure 13-24.

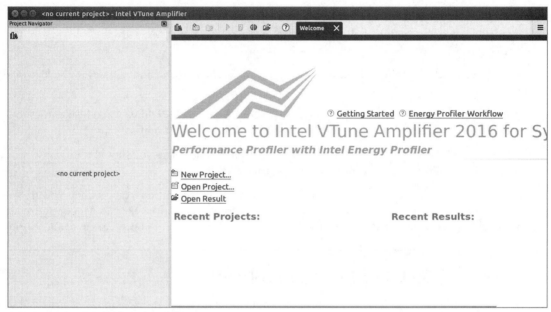

FIGURE 13-24: Intel VTune Amplifier

> **NOTE** *To use Intel VTune Amplifier, you should have an Android device with Intel SoC. x86 emulators won't work with Intel VTune Amplifier.*

To start profiling an x86 Android system, application, or process, start a new project. Just click New Project and enter a name for the project. Next, you will see a new tab next to the Welcome tab named New Amplifier Result where you choose the analysis target, as shown in Figure 13-25.

As Figure 13-25 shows, there are three targets: local, Android Device (ADB), and remote Linux (SSH). You should select Android Device to analyze your x86 Android device. You can also choose to analyze the process, system, application, or an Android package by selecting an item from the drop-down box shown in Figure 13-25, and it is also possible to select a target device just next to the target process.

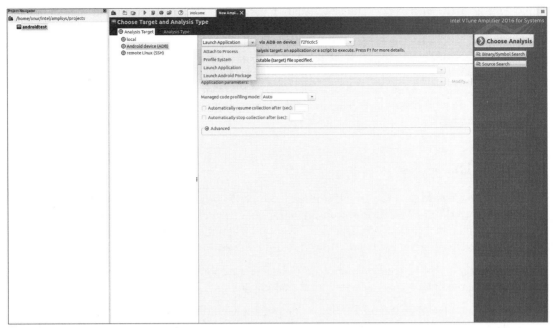

FIGURE 13-25: Intel VTune Amplifier new project view

To start collecting data from the target device, you can just click the arrow button on the toolbar. When you are finished with data collection, detailed information is displayed and you will be able to navigate between each part of the analysis data.

Intel GPA

Intel GPA (Graphics Performance Analyzers) is a set of tools like Intel VTune Amplifier but providing more high-level analysis options on Android devices. Compared to VTune, Intel GPA is less complex and easier to use. Android tools included in Intel GPA are:

➤ **System Analyzer**—This tool shows performance bottlenecks on the CPU or GPU. In order to run System Analyzer on an Intel System Studio installation, go to /opt/intel/SystemAnalyzer and run the gpa-system-analyzer-bin binary. When you run the application, you will see the list of devices available to use with System Analyzer. When you select the appropriate device and connect to it, you will see the list of compatible applications and the system option to view the performance timeline.

➤ **Platform Analyzer**—This tool provides an overview of dump data collected over time to show bottlenecks during application execution. To run Platform Analyzer, run the /opt/intel/PlatformAnalyzer/bin64/amplxe-gui binary.

➤ **Graphics Frame Analyzer of OpenGL**–Use this tool to get detailed graphics frame analysis for applications such as games that use OpenGL. To run the Graphics Frame Analyzer, go to the Intel System Studio installation path, /opt/intel/FrameAnalyzerOGL, and run the FrameAnalyzerOGL binary.

Intel GPA is available both with Intel System Studio and for standalone download and use. To download the standalone version, select Intel Graphics Performance Analyzers at `https://software.intel.com/en-us/gpa` and select Intel GPA from the right pane. To download standalone versions of tools, enter your email address, and you will be sent a link with the URL to download tools to your host OS.

INTEL INDE

Intel INDE refers to Intel Integrated Native Developer Experience with supported tools for Linux, Windows, and Android platforms for developing high performance applications running on platforms with Intel CPU and GPU. Detailed information about all the tools and libraries provided in the Intel INDE program is available at `https://software.intel.com/en-us/intel-inde`.

You must register with your email address to receive a response download links for the tools. For some tools, the links take you directly to the download URL.

Some tools are listed on the Intel INDE home page, while others such as Intel C++ Compiler are available only with Intel System Studio. Intel IPP and some others such as Intel GPA and VTune are provided both as standalone tools and with Intel System Studio. In this section, we discuss only the tools provided in the Intel INDE program that we haven't covered in the previous sections of this chapter.

Some tools in the Intel INDE that we discuss here (such as Intel Tamper Protection Toolkit and Intel Multi-OS Engine) are in beta phase. This section begins with a brief overview of Intel Tamper Protection Toolkits. Then we dive into a detailed look at Multi-OS Engine and Context Sensing SDK.

Intel Tamper Protection Toolkit

The Intel Tamper Protection Toolkit helps developers protect their application with code obfuscation and securing passwords with crypto libraries.

Like INDE tools, Intel Tamper Protection Toolkit also works with Android NDK for better protection and advanced security. More detailed information and the toolkit can be downloaded at `https://software.intel.com/en-us/tamper-protection`.

If you are developing an application with sensitive user information, or using a highly advanced and private algorithm that you want to protect against reverse engineering, Intel Tamper Toolkit can help you learn about obfuscating and securing your intellectual property and sensitive data.

Intel Multi-OS Engine

The Intel Multi-OS Engine, also provided in Intel INDE, is a framework to help developers create Android and iOS applications with Java programming languages. The Multi-OS Engine is integrated with Android Studio and XCode to generate installable binaries for Android and iOS.

The Multi-OS Engine's modules and related tools make up a single framework for creating iOS and Android applications, but it is not an easy tool to use. It is not yet a mature product, so you should expect to get errors during development. In addition, you will need to learn about Nat/J performance bindings and so on for effective application development. However, if your application will be a simple one, you are an expert with the framework, and you need to publish an application immediately on both platforms, the Multi-OS Engine can be very useful. Otherwise it can be a waste of time to learn the Multi-OS Engine framework. Continue reading if you want to see the installation instructions.

The Multi-OS Engine is available for Windows and Mac OS X; it can be downloaded at `https://software.intel.com/en-us/multi-os-engine/`.

In this example, we test the Multi-OS Engine on Mac OS X, so you should downloaded the `m_multi_os_engine-1.0.598.dmg` file.

To install Multi-OS Engine, open the `.dmg` file, then run the `Multi-OS Engine Installer 1.0.598.app`. In the first screen, the installer asks for the system password. Then it follows up with a screen asking for the direct JDK, Android Studio, and Android SDK locations, as shown in Figure 13-26.

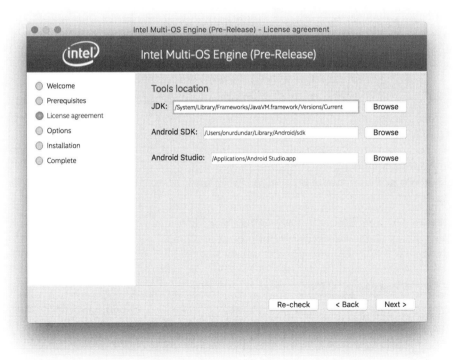

FIGURE 13-26: Multi-OS Engine SDK path selections

The installer continues with the license agreement and then the installation path selection shown in Figure 13-27.

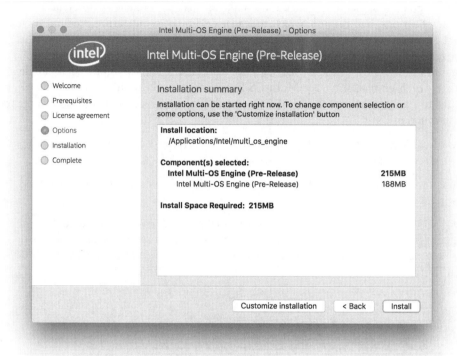

FIGURE 13-27: Multi-OS Engine Installation path selection

When the installation is done, the Android Studio plugin is installed so you can create new projects for both Android and iOS. If you reopen Android Studio and check the File ⇨ New menu, you will see that two new options are present, as shown in Figure 13-28.

When you create a new Intel Multi-OS Engine Project or Module, a new window will ask you to select the type of project or module template to create.

FIGURE 13-28: Multi-OS Engine Android Studio menu items

Intel Context Sensing SDK

The Intel Context Sensing SDK, one of the Intel INDE tools, is a free library provided by Intel for Android and Windows platforms to interact with the services and sensors on devices to create context aware applications. Intel's Context Sensing SDK can be downloaded from `https://software.intel.com/en-us/context-sensing-sdk` for Windows and Mac OS X.

Let's try it on Mac OS X to see how it works. If you download the Mac OS X version, you will get a file named `m_cssdkandroid_1.7.2.852.dmg`.

When you extract the `.dmg` file and run `m_cssdkandroid_1.7.2.852.app`, the dialogue boxes that appear ask you to install the SDK into a given directory with the traditional Intel installer interface shown in Figure 13-29.

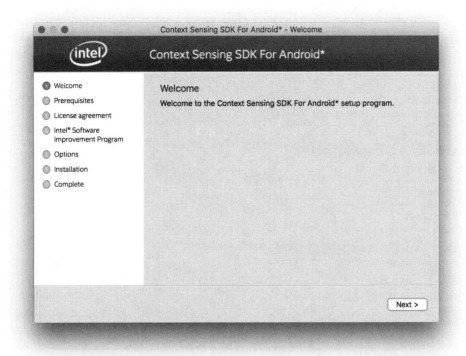

FIGURE 13-29: Intel Context Sensing SDK installer

There are multiple samples and a jar file as a library in the installation directory, as shown in Figure 13-30.

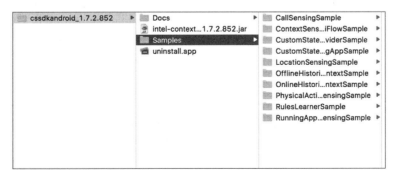

FIGURE 13-30: Intel Context Sensing SDK installation content

Let's import PhysicalActivitySensingSample. It would be good to show how the SDK works with physical sensors. During the import, there shouldn't be any errors; however, build.gradle's compile SdkVersion may be an earlier version than your installed version, so you should change it to the version you have. compileSdkVersion imports as version 8 by default.

Now you need to import the library module to your new project. Select New Module from the File menu or right-click the project. Select Import .JAR/.AAR Package from the Create New Module window. Finally, select the jar file from the installation directory, as shown in Figure 13-31, by clicking on the icon button at the right of the file name area.

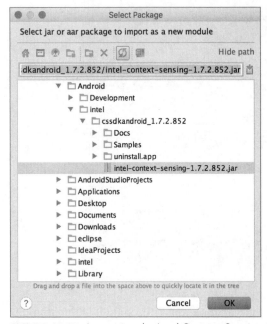

FIGURE 13-31: Importing the Intel Context Sensing SDK library

Next, you need to add the new library module as a dependency to the app module. (You learned how to add module dependencies in Chapter 7.) Right-click on the app module and select Open Module Settings. Finally, select the Dependency tab and define the dependency, or add the following gradle line to the module's build.gradle file:

```
dependencies {
    compile project(':intel-context-sensing-1.7.2.852')
}
```

Now you can run the application and see the five sensor buttons. First press 2) Start Daemon and then press 3) Enable Sensing to see toast messages on the screen about the context analysis, as shown in Figure 13-32.

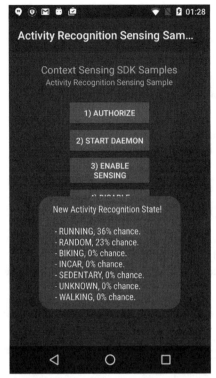

FIGURE 13-32: PhysicalActivitySensingSample screenshot

As you can see, the SDK gives lots of information about the state of the user's physical activity. Other sample applications can also be run that show the use cases and possible innovative solutions that you can add to your application without any further algorithm or data analysis.

QUALCOMM ANDROID SOFTWARE TOOLS

Qualcomm provides various software tools for Android application development for Snapdragon SoC, which Qualcomm claims is used on one billion Android devices. Like all other chip vendors, Qualcomm assists developers to enhance the performance and experience of Android applications, especially native ones, with supporting libraries and tools. Visit `https://developer.qualcomm.com/get-started/android-development` to read more about all of the tools in detail.

> **NOTE** *Qualcomm Software Tools for Android is specifically optimized for the Snapdragon CPU and GPU, so you may not observe significant improvements on tools promising performance optimizations if you ran the application on another vendor's SoC.*

Snapdragon LLVM Compiler for Android

LLVM Compiler for Android is available for Linux and Windows development platforms. LLVM Compiler for Android can be downloaded at `https://developer.qualcomm.com/software/` `snapdragon-llvm-compiler-android/tools`.

For this example, we use Snapdragon LLVM Compiler on a Linux platform to build our test applications.

Extract `snapdragon-llvm-3.7-compiler-linux64.tar` into the toolchain folder under the Android SDK installation's ndk-bundle folder. Rename the toolchains folder llvm-snapdragon-clang3.7 with the following commands:

```
$ tar xvf snapdragon-llvm-3.7-compiler-linux64.tar -C /path/to/androidsdk/
ndk-bundle/toolchains
$ mv llvm-Snapdragon_LLVM_for_Android_3.7 lvm-snapdragonclang3.7
```

As it extracts, the tar file will populate multiple folders with setup configurations, but the main build files will be contained in the `llvm-Snapdragon_LLVM_for_Android_3.7/prebuilt/linux-x86_64` folders. You can also find user guide PDF files inside the folder. These provide detailed use cases and instructions.

We will demonstrate only a basic compile and run with Snapdragon LLVM compiler using a sample application named Native Plasma.

To build your application with Snapdragon LLVM compiler, you need to add `toolchain =` `"clang"` and `toolchainVersion = "snapdragonclang3.7"` to the android.ndk definitions in the `app.gradle` file, as shown in the following code:

```
android.ndk {
    moduleName = "native-plasma"
    toolchain = "clang"
    toolchainVersion = "snapdragonclang3.7"
    CFlags.add("-I${file("src/main/jni/native_app_glue")}".toString())
    ldLibs.addAll(["m", "log","android"])
}
```

> **NOTE** *The Native Plasma application uses the experimental gradle NDK plug-in, which allows you to add* `toolchain` *and* `toolchainVersion` *properties. The stable gradle plugin will not allow you to use these properties.*

Now you are ready to build the Native Plasma application and run it. You can change the compiler for your current application immediately. For further improvements and detailed optimization options for your C/C++ code, refer to the Snapdragon LLVM compiler documentation.

Qualcomm Adreno GPU SDK

The Adreno GPU is the Snapdragon SoC's graphical processor unit. Qualcomm provides the Adreno GPU SDK to make your 2D and 3D operations better. With it, game developers can take full advantage of the GPU during development to make the final application run on the GPU as efficiently and as fast as possible.

The Adreno GPU SDK can be downloaded at `https://developer.qualcomm.com/software/adreno-gpu-sdk`. The Adreno GPU SDK can be used on Windows, Mac OS X, and Linux platforms.

The Adreno SDK download file is large and there is no direct way to use it with Android Studio. It is a pure graphics development library for developing assets with OpenGL ES 1.0, OpenGL ES 2.0, OpenGL ES 3.0, OpenCL, and DirectX (Windows phones). All source code is implemented in C and C++.

There are plenty of documents provided with the downloaded file, in the Docs folder. You can also download the OpenGL ES Developer Guide at `https://developer.qualcomm.com/software/adreno-gpu-sdk/tools`.

Qualcomm FastCV Computer Vision SDK

FastCV SDK is an image-processing SDK like OpenCV but provided by Qualcomm and optimized for Qualcomm Snapdragon SoC, and it runs on ARM-based processors. Using a library like FastCV gives you the opportunity to create applications with real time image analytics and processing from a smartphone's camera, which facilitates creating augmented reality applications with text, face, and object detection and tracking options. Implementing such operations from un-optimized libraries would consume all your battery on an Android device and take too much development time to reinvent the wheel.

More information and download links can be accessed at `https://developer.qualcomm.com/software/fastcv-sdk`. The FastCV SDK can be used on Windows, Mac OS X, and Linux platforms.

Let's integrate and test the external SDK on Android Studio for Mac OS X. The file for Mac OS X is `fastcv-installer-android-1-7-0.app`. This installer works with the Java 6 Runtime. The OS will warn you about the Java version and direct you to `https://support.apple.com/kb/DL1572` to help you install the required software after you click the More Info button on the warning popup.

Run the downloaded file and follow the instructions to complete the installation. The default installation path is `/Users/username/Android/Development/fastcv-android-1-7-0`.

When the installation is finished, the FastCV SDK, files, and libraries will be in the installation directory, as shown in Figure 13-33.

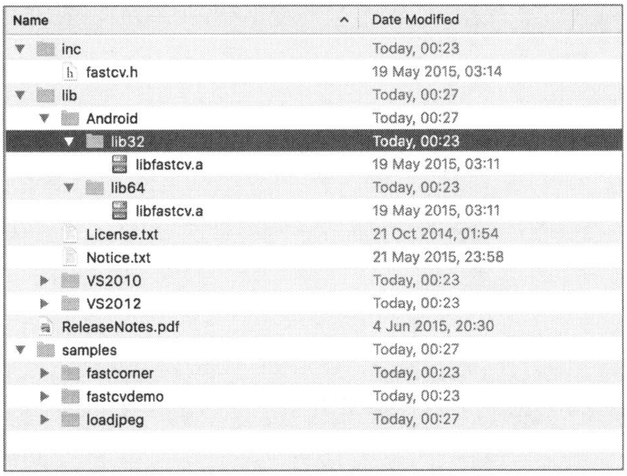

Name	Date Modified
▼ inc	Today, 00:23
h fastcv.h	19 May 2015, 03:14
▼ lib	Today, 00:27
▼ Android	Today, 00:27
▼ lib32	Today, 00:23
libfastcv.a	19 May 2015, 03:11
▼ lib64	Today, 00:23
libfastcv.a	19 May 2015, 03:11
License.txt	21 Oct 2014, 01:54
Notice.txt	21 May 2015, 23:58
▶ VS2010	Today, 00:23
▶ VS2012	Today, 00:23
ReleaseNotes.pdf	4 Jun 2015, 20:30
▼ samples	Today, 00:27
▶ fastcorner	Today, 00:23
▶ fastcvdemo	Today, 00:23
▶ loadjpeg	Today, 00:27

FIGURE 13-33: FastCV SDK directory

As you can see, there are three sample applications. Before trying to work with samples, you need to import the fastcv library header and library to Android NDK folders. If you don't import the modules to the appropriate Android NDK paths, you need to copy the files into your project folder.

Create a folder named fastcv, as shown in the following code:

```
$ mkdir /path/to/ndk-bundle/platforms/android-21/arch-arm/usr/include/fastcv

$ cp /Users/username/Android/Development/fastcv-android-1-7-0/inc/fastcv.h /path/
to/ndk-bundle/platforms/android-21/arch-arm/usr/include/fastcv

$ cp /Users/username/Android/Development/fastcv-android-1-7-0/lib/Android/lib32/
libfastcv.a /path/to/ndk-bundle/platforms/android-21/arch-arm/usr/lib
```

> **NOTE** *If you want to use the 64-bit version of the fastcv library, you can copy the header file and 64-bit library to* `arch-arm64`*'s includes and lib folders.*

Now, import the loadjpeg sample with Android Studio.

As you've done in previous chapters, you can use the Import Project option to load the sample. Android Studio will recognize the sample project as an Eclipse project and will import it. Importing native projects from Eclipse requires further actions. Because native applications use `Android.mk` files, you need to convert mk files to gradle files to make Android Studio build native code.

Importing should work without any problem and create some of the required files, as shown in Figure 13-34.

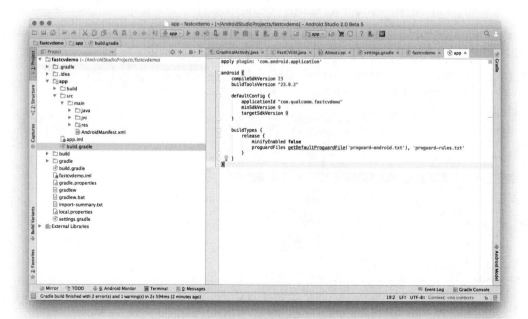

FIGURE 13-34: FastCV demo application in Android Studio

When you import the loadjpeg application, the stable gradle plugin builds the demo application, but you need to convert gradle files so they use the gradle experimental plugin to build this complex application.

> **NOTE** *If you get an error about deprecated NDK tools, you can create a* gradle.properties *file and add the line:*
>
> ```
> android.useDeprecatedNdk = true
> ```

Now you can continue to convert existing gradle files to use the gradle experimental plugin by editing the project's build.gradle file and changing classpath to the latest experimental plugin, as in the following snippet. In the current Android Studio, it is the 0.6.0-beta5 version.

```
dependencies {
    classpath 'com.android.tools.build:gradle-experimental:0.6.0-beta5'
}
```

Now, you need to edit the module's gradle file to make the sample application build the native module and load the libfastcv.a library.

Your NDK module name will be loadjpeg. Building the module will create the libloadjpeg.so file as output and a copy of the file inside the apk package during apk generation. You also need to link libraries to your native library. After that's all done, the build.gradle file will be as shown in Listing 13-1.

LISTING 13-1: loadjpeg module's build.gradle file content

```
apply plugin: 'com.android.model.application'

model {
    android {
        compileSdkVersion = 23
        buildToolsVersion = "23.0.2"

        defaultConfig.with {
            applicationId "com.qualcomm.loadjpeg"
            minSdkVersion.apiLevel = 9
            targetSdkVersion.apiLevel = 9
        }

    }
    android.ndk {
        moduleName = "loadjpeg"
ldLibs.addAll(["android", "EGL", "GLESv2", "dl", "log", "fastcv"])
        stl      = "stlport_static"
    }

    android.sources {
        main {
            jni {
```

```
                    source {
                        srcDirs 'src/main/jni'
                    }
                }
            }
        }

        android.productFlavors {
            create ("arm7") {
                ndk.abiFilters.add("armeabi-v7a")
            }
            create ("arm8") {
                ndk.abiFilters.add("arm64-v8a")
            }
            create ("x86-32") {
                ndk.abiFilters.add("x86")
            }
            // for detailed abiFilter descriptions, refer to "Supported ABIs" @
            // https://developer.android.com/ndk/guides/abis.html#sa
            // build one including all productFlavors
            create("fat")
        }
}
```

To correctly test the application, navigate to `LoadJpeg.java` and write a valid path for a JPEG file in the `onResume()` method and a correct RGB color model configuration in the `loadJPEG("jpegFilePath")` method.

Finally, you need to make sure the loadjpeg sample's Android manifest file is working. Currently, there are some errors in the `AndroidManifest.xml` file, so we corrected them as shown in Listing 13-2.

LISTING 13-2: loadjpeg AndroidManifest.xml file, corrected version

```
<?xml version="1.0" encoding="utf-8"?>
<manifest xmlns:android="http://schemas.android.com/apk/res/android"
      android:versionCode="1"
      android:versionName="1.0" package="com.qualcomm.loadjpeg">
    <application android:icon="@drawable/icon" android:label="@string/app_name">
        <activity android:name="com.qualcomm.loadjpeg.LoadJpeg"
android:label="@string/app_name">
            <intent-filter>
                <action android:name="android.intent.action.MAIN" />
                <category android:name="android.intent.category.LAUNCHER" />
            </intent-filter>
        </activity>
    </application>
    <uses-sdk android:minSdkVersion="23"  android:targetSdkVersion="23"/>
    <uses-permission android:name="android.permission.WRITE_EXTERNAL_STORAGE" />
</manifest>
```

Now you are ready to build and run the fastcv sample application. Using similar methods, you can import other samples and create your own sample project.

> **WARNING** *This library works only on ARM architecture, so you should add a fallback using other libraries for x86 and MIPS devices.*

Snapdragon SDK for Android

Snapdragon SDK for Android is built specifically for Android. It can also be used in Java code together with its native C/C++ support.

Snapdragon SDK for Android helps you develop applications with facial processing and recognition features. With the help of Snapdragon SDK, you can analyze each camera frame for blink detection, eye gaze tracking, smile, and position of head.

To start using Snapdragon SDK for Android, download the `snapdragon_sdk_2.3.1.zip` file at `https://developer.qualcomm.com/software/snapdragon-sdk-android`. Extract the zip file to create the c_cpp, java, and testapp folders together with the release and license files.

The c_cpp folder holds the shared library `.so` file used in your native application. You can find documentation of the C/C++ library in the c_cpp/docs folder.

The java folder includes docs and the Java library together with sample applications. Finally, you have a test application APK in the testapp folder.

The Snapdragon SDK test application is used to test that all promised features are working as required on your target device. You can install it with following command, using adb:

```
$ cd /path/to/android/sdk/platform-tools/
$ ./adb start-server
$ ./adb install /path/to/SnapdragonSDK/testapp/
SnapdragonSDKTestApp.apk
```

This command should install the test application. Run it on your device and see if all features pass the tests. You start the Automated Test on the initial screen. The FACIAL PROCESSING API TESTS and FACIAL RECOGNITION API TESTS buttons start the tests. When we started the FACIAL RECOGNITION API TESTS, the results looked like Figure 13-35.

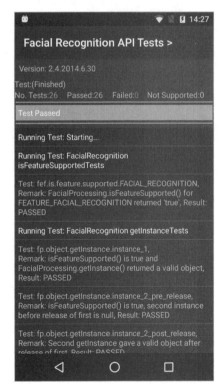

FIGURE 13-35: Facial Recognition API Tests

The test application makes sure that your device is capable of supporting the features it tests. If you see any failures, it means that you can't use the failed API feature in your application for the device you are testing.

For this example you start with a sample face recognition application from the samples directory and run it to see Snapdragon SDK's functions. This will be good practice for the techniques you learned in previous chapters.

First, import a project by clicking New ⇨ Import Project. Select the sample application under the SDK directory: /*PathTo*/SnapdragonSDK/java/samples/ samples_facial_processing/ FacialRecognitionSample.

After you have imported the sample application, note that it will not build because it is not yet linked with the library. You will see errors, as shown in Figure 13-36.

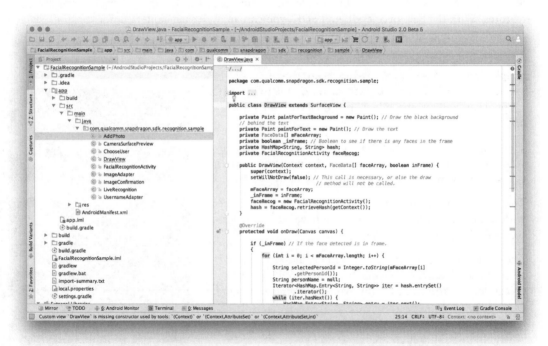

FIGURE 13-36: Importing FacialRecognitionSample

You need to import the JAR library and its references in the JNI library.

Switch to Project view in Android Studio. Create a new libs folder in the app module and copy the sd-sdk-facial-processing.jar file to the libs folder. Then right-click the jar file and click Add As Library, as shown in Figure 13-37.

After you add the jar file, Android Studio creates the gradle dependency entry, as shown in Figure 13-37. The errors have disappeared but you need a final touch to add the jar file's dependency to the so (shared object) file.

Finally, you need to add a JNI library dependency to the project. To do so, create a jniLibs folder under the src/main/ folder and copy the armeabi folder from the java/libs/libs_facial_processing folder to the jniLibs folder, as shown in Figure 13-38.

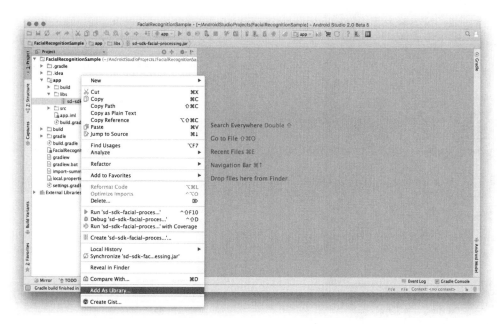

FIGURE 13-37: Adding a jar file as a library

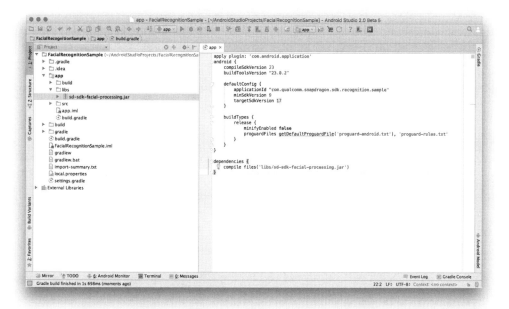

FIGURE 13-38: JNI library dependency for Snapdragon SDK

Now you can run the application on your device, as shown in Figure 13-39. You can capture images. and the face recognition sample application will recognize the faces and so on.

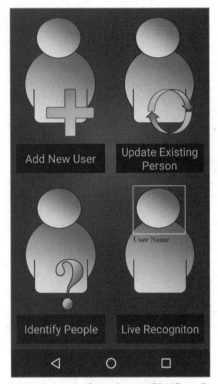

FIGURE 13-39: Snapdragon SDK Face Recognition sample application

> **WARNING** *Snapdragon SDK works only on ARM architecture devices, so it may not work on devices with x86 or mips processors.*

> **NOTE** *Further documentation and support data can be found by following the links in the left pane at* `https://developer.qualcomm.com/software/snapdragon-sdk-android`.

Qualcomm AllPlay Click SDK

Qualcomm AllPlay is a branding for the platforms that are able to play from mobile devices. AllPlay mainly exists on speakers, TVs, and similar devices to stream audio seamlessly. For devices with AllPlay enabled, the Qualcomm AllPlay Click SDK lets you integrate your application so you can stream audio to platforms supporting AllPlay. Information about supporting platforms and more can be found at `https://www.qualcomm.com/products/allplay`.

In this example, you will make your Android application compatible with AllPlay devices. Go to `https://www.qualcomm.com/products/allplay/developer-tools` and download the SDK for Android. The `allplay-click-sdk-android-v2.1.0.zip` file will be downloaded. When you extract the SDK, you'll see that the Android library is already there, so there's not much to integrating the SDK with your application.

Included with the SDK is a sample demo application together with release and debug Android libraries for use in either mode, as shown in Figure 13-40.

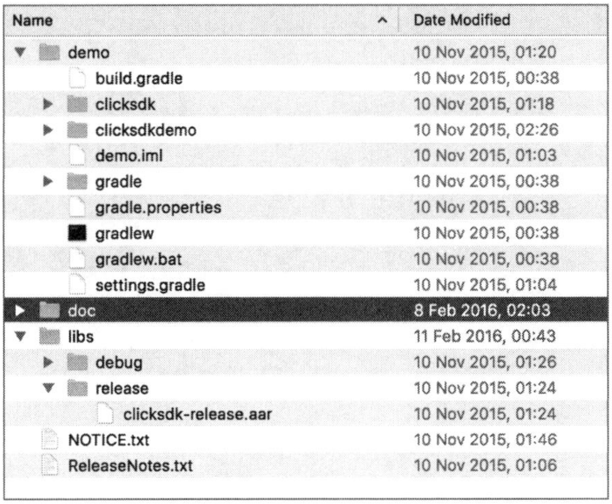

Name	^	Date Modified
▼ 📁 demo		10 Nov 2015, 01:20
📄 build.gradle		10 Nov 2015, 00:38
▶ 📁 clicksdk		10 Nov 2015, 01:18
▶ 📁 clicksdkdemo		10 Nov 2015, 02:26
📄 demo.iml		10 Nov 2015, 01:03
▶ 📁 gradle		10 Nov 2015, 00:38
📄 gradle.properties		10 Nov 2015, 00:38
⬛ gradlew		10 Nov 2015, 00:38
📄 gradlew.bat		10 Nov 2015, 00:38
📄 settings.gradle		10 Nov 2015, 01:04
▶ 📁 doc		8 Feb 2016, 02:03
▼ 📁 libs		11 Feb 2016, 00:43
▶ 📁 debug		10 Nov 2015, 01:26
▼ 📁 release		10 Nov 2015, 01:24
📄 clicksdk-release.aar		10 Nov 2015, 01:24
📄 NOTICE.txt		10 Nov 2015, 01:46
📄 ReleaseNotes.txt		10 Nov 2015, 01:06

FIGURE 13-40: AllPlay Click download package contents

To run the demo application, import it as you learned in previous chapters. It will be seamless because it is already an Android Studio project. After importing, you will see the project, as in Figure 13-41.

When you run the application, you see that it launches without any problem, as shown in Figure 13-42.

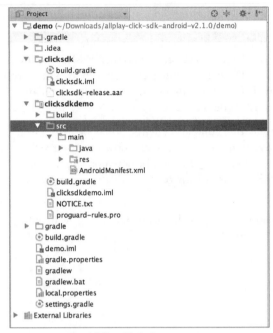

FIGURE 13-41: AllPlay Click demo project view

FIGURE 13-42: AllPlay Click demo application

If you have an AllPlay device at home, you can try it. Make sure you grant storage reading permission in the Android device by selecting Settings ⇨ Apps ⇨ Click SDK Demo and enabling Permissions.

Qualcomm Profilers

Qualcomm provides performance profilers together with supporting software libraries. There are two profilers you can use for your Android application: Adreno Profiler for GPU profiling and Snapdragon Profiler for CPU profiling. These profilers run best on Windows machines, but you can try to run them with Mono, an open source version of the .Net framework for Mac OS X and Linux.

> **TIP** *Our tests show that it is not feasible to use Mac OS X or Linux versions of profilers because the profilers are actually developed for Windows.*

> **NOTE** *The following sections do not cover the details of the profilers, only the basic installation steps to make profilers ready.*

Make sure the Android Debug Bridge executable is defined in the system path and is running before running profilers. First, add the path for adb.exe to the environment variables. The following steps are for Windows 10.

1. Right-click on This PC and then select Properties.

2. The System window will open. Click the Advanced system settings link on the left.

3. The System Properties window will launch with the Advanced tab active. Click the Environment Variables button at the bottom to open the Environment Variables window.

4. In the Environment Variables window, there will be two sections: One shows the User environment variables and the other shows the System environment variables. Select the Path variable from the System list and click the Edit button near the bottom of the window to open the list of Path variables, as shown in Figure 13-43.

5. Click New and enter the path of adb—for example, `C:\path\to\androidsdk\platform-tools\`.

    When you are done defining the path, you will see the adb command is available at the command prompt or in Power Shell. To launch adb, attach your device and run the following command:

```
$ adb start-service
```

> **WARNING** *If you selected the Linux or Mac OS X version of adb, make sure the adb executable is defined in the system PATH variable and your device is attached to adb.*

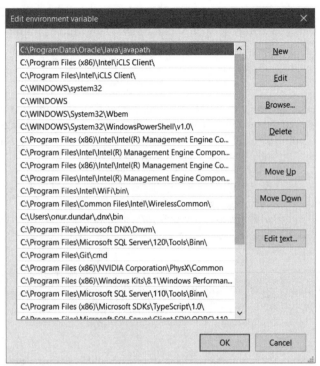

FIGURE 13-43: Windows system path variables list

Adreno Profiler

Adreno Profiler is used for GPU profiling of Snapdragon SoC platforms. This profiler can be downloaded at `https://developer.qualcomm.com/software/adreno-gpu-profiler`. Select the version you need: Windows, Max OS X, or Linux.

> **WARNING** *Adreno Profiler only works with devices having Adreno GPU– and Adreno Profiler–enabled devices, not with any others.*

Snapdragon Profiler

Snapdragon Profiler is used for CPU profiling Snapdragron SoCs. For Windows, Mac OS X, and Linux platforms, the Profiler can be downloaded from `https://developer.qualcomm.com/software/snapdragon-profiler`. The example here uses the Windows version, `snapdragon profiler-windows-1-3.zip`.

Extract `snapdragonprofiler-windows-1-3.zip` to the folder of your choice. Before running the `SnapdragonProfilerSetup.exe` file, make sure GTK# (the Windows version of Gtk) is installed.

If it is not, the Snapdragon installation will direct you to the following URL to download the GTK# installer: `http://download.xamarin.com/GTKforWindows/Windows/gtk-sharp-2.12.25.msi`.

When you finish installing GTK#, run `SnapdragonProfilerSetup.exe` and complete the installation.

If you set the adb path and started the adb service, you can navigate to the Snapdragon Profiler path and run the profiler as shown in the following command.

```
$ cd C:\Program Files (x86)\Qualcomm\Snapdragon Profiler
$ .\SnapdragonProfiler.exe
```

If nothing happens, select File ⇨ Connect and wait for your device to be connected. It usually takes a couple of seconds and you are ready to use Snapdragon Profiler, as Figure 13-44 shows.

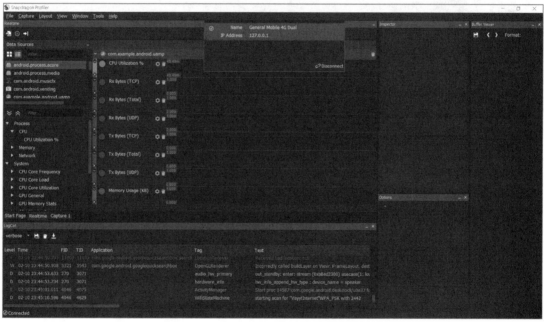

FIGURE 13-44: Snapdragon Profiler in use

NVIDIA SOFTWARE TOOLS

NVIDIA provides CodeWorks software tools for Android application developers. NVIDIA CodeWorks is available for all Windows, Mac OS X, and Ubuntu 32- and 64-bit versions. You can download CodeWorks at `https://developer.nvidia.com/codeworks-android`.

Click the download link to get an executable to install the required tools and libraries onto your development machine. (`CodeWorksforAndroid-1R4-linux-x64.run` for Ubuntu, `CodeWorksforAndroid-1R4-osx.dmg` or `CodeWorksforAndroid-1R4.windows.exe`).

You can run the Ubuntu version with the following commands:

```
$ chmod +x CodeWorksforAndroid-1R4-linux-x64.run
$ ./CodeWorksforAndroid-1R4-linux-x64.run
```

After the installer starts, select the download directory on the third screen, as shown in Figure 13-45.

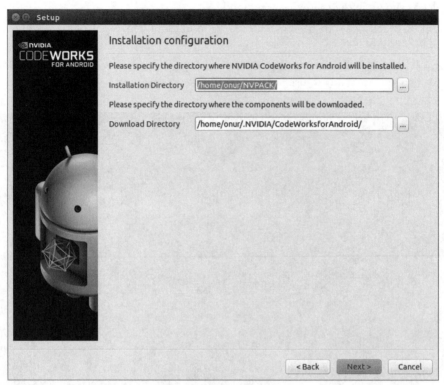

FIGURE 13-45: NVIDIA CodeWorks path configuration

When you finish configuring the path, the component configuration window opens, as shown in Figure 13-46. Here you can select the tools and libraries to download and install on your development machine.

Under Android SDK, you can find many SDK versions to select for installation. We selected only Marshmallow because we want to cover only NVIDIA's own tools, such as PerfKit, System Profilers, CUDA, and so on.

NVIDIA's focus is mostly on the GPU side to enhance use of its Tegra GPUs with CUDA, PhysX, and OpenCV. These help you make use of the GPU for more than just its normal tasks. To test and use these libraries, you should get a device with a Tegra GPU on it.

If you are going to support NVIDA Tegra or you just want to make sure your high-end game will run best on Tegra, you can try the tools as documented.

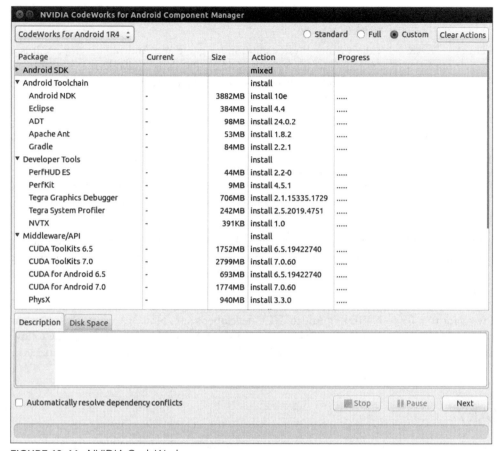

FIGURE 13-46: NVIDIA CodeWorks components

There is no support for Android Studio yet, only Eclipse as shown in Figure 13-46. The CodeWorks installer downloads a copy of Eclipse with all installations.

NVIDIA CodeWorks provides the following tools and libraries:

➤ **CUDA**—Nvidia's famous CUDA development tools are also available for Android platforms. When installed, CUDA tools and samples will be populated in the installation directory. More information about CUDA for Android can be found at `http://docs.nvidia.com/ gameworks/index.html#technologies/mobile/cuda_android_main.htm%3FTocPath% 3DTechnologies%7CMobile%2520Technologies%7CCUDA%25C2%25A0for%2520Android %7C_____0.`

➤ **OpenCV for Tegra**—The OpenCV library is also provided with the CodeWorks installation. OpenCV libraries are optimized for Tegra GPUs; CUDA acceleration can also be used with OpenCV functions. More information about OpenCV on Tegra can be found at `http://docs.nvidia.com/gameworks/index.html#technologies/mobile/`

`opencv_main.htm%3FTocPath%3DTechnologies%7CMobile%2520Technologies%7COpenC`
`V%2520for%2520Tegra%7C_____0.`

➤ **PhysX**—PhysX is Nvidia's game physics solution for game and Android developers. It is a very complex SDK. More information about Nvidia's PhysX is available at `http://docs` `.nvidia.com/gameworks/index.html#gameworkslibrary/physx/physx.htm%3FTocPath` `%3DGameWorks%2520Library%7CPhysX%7C_____0.`

➤ **NVIDIA Profilers**—Like Intel and Qualcomm, NVIDA also provides its own profilers: Tegra System Profiler and PerfHUD ES (Profiling OpenGL ES code). These profilers are more useful for embedded developers and high-end game developers. Basic use cases are available at `http://docs.nvidia.com/gameworks/index.html#technologies/mobile/native_` `android_profiling.htm%3FTocPath%3DTechnologies%7CMobile%2520Technologies` `%7CNative%2520Development%2520on%2520NVIDIA%25C2%25A0Android%2520Devices` `%7C_____5.`

➤ **NVIDIA Shield**—NVIDIA Shield is a gaming platform for consumers running the Android operating system. Its main purpose is to show Tegra's performance and provide a game console. You can visit following URL to learn more about NVIDIA Shield `https://shield` `.nvidia.com/`

Finally, we want to refer to NVIDIA's Android TV developer's site. NVIDIA wants their hardware in more Android TV devices and to increase their support of Android TV developers. The following URL provides plenty of information for Android TV developers: `https://developer.nvidia.com/` `android-tv-developer-guide.`

SUMMARY

This chapter covered some useful plugins and third-party tools and software libraries that you can use with Android Studio and SDK. We also covered supporting tools that work with Android Studio to power your development process and help you create better applications.

We started with some popular Android plugins and their external extensions. That was followed by discussion of Android SoC vendor tools and libraries, which can help you optimize your applications' power and performance. Those tools and libraries are provided by Intel, Qualcomm, and NVIDIA.

INDEX